Harvard Dissertations in the HISTORY of SCIENCE

Edited by
Owen Gingerich

Harvard-Smithsonian Center for Astrophysics

A GARLAND SERIES

Darwinism and Its Data

The Adaptive Coloration of Animals

Muriel L. Blaisdell

Garland Publishing, Inc.
New York & London
1992

Library of Congress Cataloging-in-Publication Data

Blaisdell, Muriel L. (Muriel Louise)
Darwinism and its data : the adaptive coloration of animals /
Muriel L. Blaisdell.
p. cm. — (Harvard dissertations in the history of science)
Includes bibliographical references.
ISBN 0-8240-7253-7 (alk. paper)
1. Color of animals. 2. Protective coloration (Biology) 3. Evolution (Biology) I. Series.
QL787.B53 1992
591.57'2—dc20 91-47029

Printed on acid-free, 250-year-life paper.
Manufactured in the United States of America

Darwinism and Its Data

The Adaptive Coloration of Animals

Acknowledgements

I am glad for the opportunity to thank the many people whose ideas and good-will have made it possible for me to prepare this dissertation. At the outset I would like to thank members of the Department of General Science at Oregon State University for introducing me to the study of the history of science. I am especially grateful to Paul Lawrence Farber, Margaret J. Osler, and Robert James Morris for encouraging me to continue my studies in the history of biology.

The insights and ideas of each of my professors at Harvard have surely found their way into this dissertation. I am particularly indebted to Professors Joan Cadden, Donald Fleming, Erwin Hiebert and Barbara Rosenkrantz, whose courses and positive examples have been helpful to me. I would like to offer my special gratitude to Professor Ernst Mayr, who has been extremely generous to me with both his ideas and his time. I am thankful to Professor Stephen J. Gould for making important comments and corrections in a draft of this dissertation. I am also grateful to Professor Everett Mendelsohn for his counsel during my graduate career and for his efforts in the supervision of this dissertation.

This is perhaps my final opportunity in a formal way to thank the office staff of the Department of the History of Science for their important contributions to the happiness and general welfare of the graduate students in the department--my thanks to the inhabitants of Holyoke Center 838, Elizabeth A. C. Smith, Eunice Good and Ruth Bartholomew. Also the staff of the Library of the Museum of Comparative Zoology at Harvard University has been uniformly friendly and helpful to me in my research.

I acknowledge gladly the moral support and ideas of my friends and colleagues, particularly Charlotte Porter, Molly Noonan, Joseph Chillington, Frank Sulloway, David Kohn, Leslie Burlingame, Jean Christensen and Robert McCormick. Karen Reeds has read several chapters of this dissertation and offered insightful comments and suggestions. Shirley Roe, who was also writing a dissertation this year, has been a true comrade during this last and greatest effort of our careers as graduate students. Roberta Beeson deserves far more thanks than a brief acknowledgement can convey since she listened to most of the ideas in this dissertation first and many times.

A Harvard University Fellowship provided financial support for this final year of my graduate work and I am particularly grateful to Dean Peter McKinney for making it possible.

Priscilla Braids Bellairs not only typed this dissertation, but substantially improved it.

Finally, I wish to express my deepest gratitude to

my parents, Roger S. and Bertha R. Blaisdell, for their unfailing patience and constant encouragement. This dissertation is dedicated to them.

TABLE OF CONTENTS

LIST OF ILLUSTRATIONS

PLATE I

PIERIDAE OF DOUBLEDAY AND WESTWOOD, 1846-1850

The butterfly on the lower left, identified as Leptalis Eunoe, was catalogued correctly within the family Pieridae despite its stronger resemblance to the Heliconiidae (see Plate II, p. 78). (From E. Doubleday and J. O. Westwood, The Genera of Diurnal Lepidoptera, 1846-1850, Plate V).

INTRODUCTION

The history of the theory of evolution by natural selection is now more accessible to research than ever before. Several recent publications have presented sources formerly found only in manuscript collections in England. The publication of Darwin's notebooks on the transmutation of species and on psychology have been particularly important; and the publication of a complete edition of Darwin's letters is in prospect and eagerly awaited.[1] The resurgence of interest in the history of evolution sparked in large proportion by the centennial publications of 1959 has increased steadily to the present day.

Although historians of evolution theory have begun their efforts at the proper center, that is, with Darwin himself, a good body of literature has also begun to accumulate which treats the labors of the other early proponents of evolution. Darwin as an individual has sometimes been confused with Darwinism; not all the ideas justifiably called Darwinian are Darwin's own. Some histories of the evolutionary theory have been flawed by excess partisanship in the attempt to discern Darwin's contributions to science in opposition to those of his immediate circle of colleagues. Historical studies do not require heroes and villains in the way that drama oftentimes does.

There is one problem for which Darwin was not the most important explicator, that of the relationship between animal colors and evolutionary theory. To say this ought not to be construed as a denigration of Darwin's important contributions. Too much historical work has already been written which has either a pro- or anti-Darwin partisan quality. For example, C. D. Darlington accused Darwin of completely muddling the evolutionary issues so that his views might be favorably received by naturalists representing very different points of view on the sources and directedness of variation.[2] The verbal portraits of Darwin drawn by Gertrude Himmelfarb, William Irvine, and Jacques Barzun tend to be ungenerous in their characterizations.[3] The otherwise excellent study of Darwin in Michael Ghiselin's <u>The Triumph of the Darwinian Method</u> seems to me to err in the opposite direction of conflating Darwin's personal contributions with those of other Darwinians who had vital parts in the theoretical development, data collection, and popularization of evolution by natural selection.[4] Instead of entering into the turbulent area of the history of science that deals specifically with Darwin as an individual, this thesis instead presents a study of one class of data, the coloration of animals, and assesses the part it played in the Darwinian controversy.

One reason for subjugating the roles of particular

personalities to the informational content of scientific papers and books was my increasing awareness that Charles Darwin, Alfred Russel Wallace, and Henry Walter Bates were never the only naturalists involved in the theorizing, experiment making, or observing. The remaking of the biological sciences in the nineteenth century involved many people and included many aspects both internal and external to science.

This research project began as a study comparing the writings and theories of Darwin and Wallace. As I proceeded to follow their references to the work of their predecessors and contemporaries, it became increasingly clear that the account of the relationship between the data class (adaptive coloring in animals) and the evolutionary theory, could only be told properly by not isolating the work of a few chief Darwinians from that of a wider community of scientific workers.

Animal colors and their adaptive significance had long been a subject that interested nearly every naturalist, from the humblest collector to the professionals at the British Museum. The diverse adaptations of organisms was one subject that could be almost certain to generate animated discussion whether the naturalists present were committed to a view that adaptation was implanted by a benevolent Creator or produced by natural processes. Representatives of opposite positions on metaphysical questions shared a common interest in adaptation.

Although Henry Walter Bates was first and Wallace second to assert that mimicry in butterflies was not a result of creation but was slowly developed through the process of natural selection, they were supported in their claim by other naturalists. In the 1860's and 1870's the Darwinian side of the argument was taken by naturalists who are scarcely noted at all in the standard encyclopedia and dictionaries of biography. Raphael Meldola, A. G. Butler, J. Jenner Weir, and R. Trimen, for example, are often not accorded even a modest niche in the pantheon of Darwinian heroes. Nevertheless, they were part of the positive reception of Darwin's theory, and they published their support when it mattered very much. The role of the "non-elite" Darwinian scientists has been little studied in terms of their influence upon the gradual process of teaching scientists to think about species in a new way.

In his Darwin and the General Reader, Alvar Ellegård expounds his findings that the response to Darwinian evolutionary ideas was very general, spreading throughout the British press from the highbrow journals to the lowbrow penny magazines.[5] Of course there were important differences in the reception of Darwinian ideas at each stratum. The use of religious rebuttals, for example, varied. The presence of some form of response was quite general, however. Studies of the reception of Darwinian theory in the scientific press also show that response was universal in its presence, although heterogeneous in its form. The lower prestige journals of societies and serial publications were often more forthright than "highbrow" journals in their confrontations of Darwinian subjects.[6] The higher prestige journals, on the contrary, published articles with obvious bearing on the subject of evolutionary change but devoid of any explicit mention of evolution or of natural selection.

The Entomological Society of London has been a particularly interesting group to study because it chose a middle road between the introduction of non-scientific declamations of evolutionary theory and the form of tacit argumentation epitomized by the Royal Society's _Transactions_. Discussions of evolution were virtually insured by the attendance at meetings of both Bates and Wallace. Explicit statements for or against evolutionary theory may be found both in the published _Transactions_ and in the proceedings of the meetings. The subject of adaptive coloration was discussed at the Entomological Society for many years before the _Origin of Species_. Given the predilections of society members for taxonomy and collecting, they often perceived the Darwinian claims to be able to explain the external form and color of organisms as a challenge to the orderly conduct of the science. The bearing of evolutionary theory on the practice of entomology was too obvious to be discussed euphemistically or politely disregarded. Positions were taken and staunchly defended. The subject of animal colors and the theoretical framework explaining color was a test case for the Darwinian theory. Naturalists who used the design arguments of natural theology engaged the Darwinians in prolonged and serious debate. The partisans of both sides felt that here was a concrete topic which could be used to test the claims of the Darwinian theorists. It seemed that natural selection could be proven either false or true, once and for all time, by careful, systematic judgment of how well it accounted for the phenomena of animal form and color.

The Entomological Society of London was the single most active group in presenting arguments about theories and offering data. None of the common metaphors typifying scientific change works well in characterizing the changes wrought by the Darwinian theory for entomology. The metaphors comparing change in science to political revolution or to warfare and ultimate victory on a battlefield both fail to express indecisive qualities in the Darwinian debates. The richer mélange of metaphors and categories now available in which to think about scientific change is a hopeful thing in the history of science, however.

In his criticism of the use of Kuhnian concepts of scientific revolution through paradigm changes, John Greene has indicated that in the Darwinian "revolution" the old paradigms associated with natural theology are never quite vanquished.[7] In natural history old paradigms are given a new lease on life by effecting compromises with the new. No _coup d'état_ topples the _ancien régime_. It depends upon which elements of the Darwinian debate one stresses whether the continuity or discontinuity of biological science emerges as the dominant feature. The Darwinian revolution is notable for its incompleteness in the nineteenth century. The concept of evolution found its way into science and society at least seven decades before its most important Darwinian component, natural selection, became accepted as an important part of evolutionary theory in actual practice. The picture of scientific change as a revolution does not hold up very well with respect to the Darwinian controversy since it is a revolution without a termination and a battle without a _coup de grace_. The reduction of the complex knot of historical components that surrounds the reception of Darwinian theory has not yet been particularly successful. Historical model-building remains without rigor; like fingerprints, historical episodes have much in common from case to case, but in the last analysis no two are quite the same.

Few students of the history of biology, however, decline to name the years following the publication of the _Origin of Species_ as revolutionary--meaning a period marked by profound disturbance of the status quo and notable for the introduction of new ideas. Even so, historians have searched the Darwinian literature for roots and heritage in the science and social writings of the period before 1859. These scholars have found many evolutionists before Darwin and generally agree on the conclusion that the idea of gradual change was in the culture if not "in the air" in the early Victorian period.[8] The predecessors of Darwin have been named, by some, the "forerunners" of Darwin.[9] The history of biology, however, does not lead linearly toward Darwin through the work of a chain of forerunners whose work dimly portends a coming greater revelation. Walter Cannon's study of the natural theologians of the early nineteenth century has indicated that Darwin owed more to the natural theologians, in terms of both data and perspective, than to the speculative evolutionists.[10] Frederick Churchhill once said in a lecture that "History abhors a straight line, just as nature abhors a vacuum." It is quite certain that the history of ideas is nothing like a line.

The reception of Darwinian theory is not exactly like a political revolution and the evolutionists before Darwin are not really his "forerunners." Finally, his associates who became convinced that evolution had, in fact, occurred are not very much like "converts." Likening the acceptance of Darwinism to religious conversion does not capture the essence of the meaning of recruitment to the Darwinian position. Darwin used the metaphor of conversion to describe Lyell's declaration of the reality of evolutionary change.[11] It is likely that some people did see the adoption of the evolutionary theory as the taking of a vow or adopting a creed. For most of the individuals studied in this dissertation, the term "conversion" is too strong. Conversion connotes a whole-hearted acceptance of a new world-view, involving a radical overhaul of fundamental assumptions. During the early years of the Darwinian controversy, many--in fact most--naturalists accepted a compromise view that allowed them to admit species change without abandoning the basic tenets of older scientific belief or procedure. The members of the Entomological Society, for example, were too often only partly "converted"; they admitted a kind of evolution shorn of its mechanism, natural selection. That's like a missionary claiming as converts natives who accept one and one-half persons of the Trinity. No one metaphor, model, or simile can exhaust the complexity of the Darwinian debates; if it could the events would not be nearly so compelling and interesting as they are.

The reasons why some people accepted Darwinian evolution, others compromised and still others rejected the hypothesis must be sought by including many factors of biography, intellectual and social history. After all, ideas do not behave, but people do. In studying the idea of evolution by natural selection, it has been necessary to avoid unguarded allegiance to the history of ideas approach as against an approach to the history of science that focusses only on social factors impinging on science. The decision to include materials both internal and external to science was mandated by the selection of a study which emphasized a larger number of people. Also, the combination of social and ideational factors seems more obvious in studying the middle or lower ranks of scientific workers than in studying the revolution-makers. Probably that is because it is easier to identify with the Entomological Society Fellows than with Copernicus, Galileo, Newton, Einstein and Darwin.

This dissertation has afforded the opportunity to employ a variety of historiographical techniques. The history of ideas approach, exemplified by the *Great Chain of Being* by Arthur O. Lovejoy and *The History of Ideas* by George Boas, has been most useful to me in attempting to track the concept of evolution by natural selection through the thickets of Victorian scientific prose.[12] Articles, mostly of high quality, on the general history of Victorian England and on the religious trends and philosophy of the period contained in the journal *Victorian Studies* have added considerably to the elucidation of the context into which the Darwinian theory was introduced.[13] The efforts of both intellectual and social historians have been utilized in the pursuit of this study of the interaction of data and theory and reception of Darwin's views in the nineteenth century. The history of science may be studied by either *in vivo* or *in vitro* methods, but if a combination of methods can be used, so much the better.

Adaptive coloration in animals, particularly mimetic resemblances, has been the subject of intense study in zoology from the nineteenth century down to the present time. Wolfgang Wickler, an expert on animal coloration, has written that 1500 papers have been published on the subject of mimicry alone in the one hundred years following Bates's paper on mimicry.[14] That represents a rate of more than one paper each month for a century. The subject of animal colors, therefore, has been large enough and active enough to merit book length works of which the *Adaptive Coloration in Animals* by Hugh P. Cott is still best.[15] His bibliography, containing 685 entries, includes most of the important contributions to the field up to 1940.

The relation of the data of animal color to evolutionary theory from an historical viewpoint has also been broached by several people. Wilma George, for example, in her biography of Alfred Russel Wallace, discusses mimicry and Wallace's contributions to the confirmation of Bates's Darwinian analysis of the analogous similarities between species.[16] Her book, *Biologist Philosopher*, emphasizes the opposition between Wallace and Darwin on the subject of animal color and its development by natural selection or by sexual selection. Wilma George wrote as a scientific biographer of Wallace; her perspective is complemented, therefore, by Mary Alice Evans's treatment of the relationship of Darwinian evolution theory and the data afforded by mimetic animals.[17] Her article "Mimicry and the Darwinian Heritage" clearly invited further research, both scientific and historical. She concluded:

> The manner in which adaptations arise, or are lost, through natural selection offers a never-ending source of study to biologists. Mimicry, although it is only one of countless examples, provides a unique opportunity for the study of most of the questions which might arise in connection with the general subject of adaptations.[18]

That mimicry provides this unique opportunity is traceable to interest in animal colors and analogous resemblances that find roots even before the _Origin of Species_ and Bates's paper on the Brazilian Heliconiidae. The Darwinian reappraisal of the meaning of adaptations was argued out with reference to particular empirical data. Mimicry and other adaptive coloration phenomena provided the data for many of the debates.

Notes

INTRODUCTION

1. Sir Gavin DeBeer, editor. "Darwin's Journal," _Bulletin of the British Museum (Natural History) Historical Series._ 2 (1), 1959; "Darwin's Notebooks on the Transmutation of Species. First Notebook. (July 1837-February 1838)" 2 (2), 1960; "Second Notebook. (February to July 1838)" 2 (3), 1960; "Third Notebook. (July 15, 1838-October 2, 1838" 2 (4), 1960; "Fourth Notebook (October 1838-July 10, 1839)" 2 (5), 1960; and "Pages Excised by Darwin," 3 (5), 1967.

Howard E. Gruber, _Darwin on Man: A psychological study of scientific creativity._ With Darwin's early and unpublished notebooks, transcribed and annotated by Paul H. Barrett.(New York: Dutton, 1974).

The Darwin letters are to be published with the sponsorship of the American Council of Learned Societies and with the support of the National Endowment for the Humanities and the National Science Foundation.

2. C. D. Darlington, "The Origin of Darwinism," _Scientific American_ 200:60-66, 1956.

3. Gertrude Himmelfarb, _Darwin and the Darwinian Revolution._ The Norton Library (New York: W. W. Norton, 1968) and _Victorian Minds._ Harper Torchbooks (New York: Harper & Row, 1970). William Irvine, _Apes, Angels, and Victorians._ Meridian Books (New York: World Publishing Co., 1970). Jacques Barzun, _Darwin, Marx, Wagner_ (2nd ed.: Garden City, New Jersey: Doubleday & Co., 1958).

4. Michael T. Ghiselin, _The Triumph of the Darwinian Method_ (Berkeley: University of California Press, 1969).

5. Alvar Ellegård, _Darwin and the General Reader. The Reception of Darwin's Theory of Evolution in the British Periodical Press, 1859-1872._ Gothenburg Studies in English VIII. Göteburg, 1958.

6. Frederick Burkhardt, "England and Scotland. The Learned Societies," pp. 32-74 in _The Comparative Reception of Darwinism,_ Thomas Glick, editor. (Austin: University of Texas Press, 1974).

7. John Greene, "The Kuhnian Paradigm and the Darwinian Revolution in Natural History," in _Perspectives in the History of Science and Technology,_ edited by Duane H. D. Roller (Norman: Oklahoma University Press, 1971).

8. Milton Millhauser, Just Before Darwin: Robert Chambers and Vestiges. (Middletown, Connecticut: Wesleyan University Press, 1959).

9. Bentley Glass, Owsei Temkin, William L. Straus, Jr., editors. Forerunners of Darwin: 1745-1859 (Baltimore; Johns Hopkins University Press, 1959).

10. Walter B. Cannon, "The Bases of Darwin's Achievement: A Revaluation." Victorian Studies 5:109-134, 1961.

11. Charles Darwin to Asa Gray, September 26, 1860. The Life and Letters of Charles Darwin, edited by Francis Darwin. 3 vols. (London: John Murray, 1888) II, p. 345.

12. Arthur O. Lovejoy, The Great Chain of Being. A Study of the History of an Idea. Harper Torchbooks. (New York: Harper & Row, 1960). George Boas, The History of Ideas. An Introduction (New York: Charles Scribner's Sons, 1969).

13. Some examples are: Donald Fleming, "Charles Darwin, the Anaesthetic Man," Victorian Studies 4:219-234, 1961. John Passmore, "Darwin's Impact on British Metaphysics," Victorian Studies 3:19-40, 1959. J. W. Burrow, "Evolution and Anthropology in the 1860's: the Anthropological Society of London, 1863-71," Victorian Studies 7:139-154, 1963.

14. Wolfgang Wickler, Mimicry in Plants and Animals. Translated from German by R. D. Martin. World University Library (New York: McGraw-Hill Book Company, 1968) p. 13.

15. Hugh B. Cott, Adaptive Coloration in Animals. (London: Methuen & Co., 1940).

16. Wilma George, Biologist Philosopher. A Study of the Life and Writings of Alfred Russel Wallace (New York: Abelard-Schuman, 1964).

17. Mary Alice Evans, "Mimicry and the Darwinian Heritage," Journal of the History of Ideas 26:211-220, 1965. Also see Jeanne E. Remington and Charles L. Remington, "Mimicry, a Test of Evolutionary Theory," Yale Scientific Magazine, October, 1957, pp. 10-21, and Charles L. Remington, "Historical Backgrounds of Mimicry," Proceedings of the XVI International Congress of Zoology IV:143-149, 1963.

18. Evans, "Mimicry and the Darwinian Heritage," p. 220.

Chapter I

NATURAL THEOLOGY AND NATURE'S DISGUISES

Design Argument

Henry Walter Bates's paper on the phenomenon of mimicry in butterflies was read at the Linnean Society of London on November 21, 1861, three days short of the second anniversary of the publication of the *Origin of Species*.[1] In his paper Bates gave the details for the surprising resemblances between butterflies from different families. He explained their similarities of coloration and pattern on what he called "Mr. Darwin's principles." Bates had adopted both the general principle of evolution and the mechanism of natural selection. Without mincing his words, he wrote, "I believe the case offers a most beautiful proof of the truth of the theory of natural selection. It also shows that a new adaptation, or the formation of a new species, is not affected by great and sudden change, but by numerous small steps of natural variation and selection."[2] Bates held that natural selection had long been in operation, gradually perfecting the deceptive resemblance of *Leptalis* butterflies, which were readily eaten by many bird predators, to models of a different family, the Heliconiidae, which were distasteful to most birds. Unconvincing models were eaten by the multitude of ravenous Amazonian birds. Making it clear that the beautiful analogy of appearance was not the result of the handiwork of the Designer, or a contrivance in which the Contriver might be dimly perceived, Bates asserted forthrightly that the impression that some innate principle caused the analogy was wrong and that it was actually the work of identifiable selecting agents. The similarity in external colors between the Leptalis and Heliconiidae was a product of natural selection carried out by birds. It was an illusion to believe that these analogous resemblances were in any sense predestined.[3]

Bates's work on mimicry was an innovation for science in the sense that it was the announcement of an abrupt change in the connection between data and explanation. His use of the Darwinian mechanism of natural selection was new and very controversial. The basic concept of organisms having analogous resemblances to each other was not at all new, however. Even the term, "mimicry," was already familiar in connection with analogous resemblances. When it is said that Bates discovered mimicry it ought not to be assumed that he was the first to see the uncanny similarity between insects in quite disparate taxonomic categories. In modern works, if history is discussed at all, the subject is taken up with Bates's paper at the Linnean Society as if the phenomenon had been brought forward for the first time. That is not strictly true, although Bates did make a new subject of mimicry which is still an active area of research in entomology today. Although Bates was not the first to discuss mimetic analogies he was the first to do so using Darwinian principles. Bates's study signals one of the first, and most provocative, instances in which data taken from field observation were held to verify Darwinian expectations.

Several historians of the reception of Darwinian theory have correctly stated that the battles over evolution and natural selection were both prolonged and heated because the Designer posited by natural theologians was replaced by a natural and not necessarily directional mechanism.[4] Although it was possible to reconcile evolution to the argument from Design, as Asa Gray quickly saw, one could only do so by adjusting one's conception of the relationship between the Supernatural and the Natural to accomodate it.[5] It has been pointed out, principally by Walter Cannon, that Darwin and his earliest followers borrowed heavily from the style as well as the data of their predecessors who considered themselves equally as natural theologians and as natural historians.[6] Generally, the history of the evolutionary theory has been studied with an eye toward the discovery of discontinuities before and after 1859, or with the hope of finding so many "forerunners" that there appears to be no change at all. On the basis of this investigation it seems clear that there was a sharp and major theoretical change.

However, with respect to the empirical data used to support the new Darwinian assertions, there was much continuity with the data already in use to support the argument from design.

The coloration of animals was a favorite subject with natural theologians and had been for many years before 1859 or 1861. Animal colors were used to exemplify the concern of Providence for all creatures, but that was only the most obvious and not the exclusive use for knowledge of external morphology and adaptive function of color in animals. The natural theologians worked long and arduously toward a complete explication of the text of nature. They knew nature intimately, both from field study and from taxonomic study of very good collections. The difficulty Bates experienced in his attempts to convince his fellow naturalists of the truth of his interpretation of mimetic data was due to its familiarity to naturalists rather than to its newness. The reinterpretation of familiar data seems psychologically much more difficult to handle than the problem of grasping a new explanation in connection with a new phenomenon. Just as Darwin's Origin of Species was an attack on the current theistic version of Design, Bates's study of mimetic analogies in insects was a specialized version of Darwin's general attack upon design. The subjects of animal coloration and protection as well as the analogous resemblances between unrelated species was ground that had long been held by the natural theologians and would not be easily yielded to the Darwinian proponents.

Bridgewater Treatises

The importance of the subject of protective coloration and analogous resemblance is illustrated by the emphasis placed on it in the Bridgewater Treatises. Each Treatise which contains a discussion of animals and their adaptations describes the surprising likeness frequently observed among animals which are not structurally related.[7] Peter Mark Roget, William Kirby, and William Prout developed the subjects of analogous resemblance, protective coloration, and beauty as evidence of design in volumes V, VII, and VIII of the Bridgewater series.[8] The existence and attributes of the Creator seemed clearly manifested in the repeated colors and forms of the creatures, and in their adaptive and arrangement. The most elaborate exposition on animals was by the Reverend William Kirby, rector of Barham, and the first president of the Entomological Society of London. His two volume treatise was published in 1835 under the title, On the Power, Wisdom and Goodness of God as Manifested in the Creation of Animals and in their History, Habits and Instincts.[9]

Kirby was the ideal candidate to write a book which combined natural history with theology. He was a firm believer in the thesis that within the revelation of Scripture, particularly that of the Old Testament, a full natural philosophy lay undiscovered. He interpreted the Bible symbolically rather than literally, believing that the revelation in the Hebrew language contained the necessary materials for understanding the message of the New Testament as well as a perfect natural philosophy.[10] The precedent for attention to analogous representations and disguises in nature, Kirby found, had been established by Solomon in the Proverbs: "It is the glory of God to conceal a thing: but the honour of kings is to search out a matter."[11] Throughout his treatise Kirby described animal behavior by tightly connecting observations of animals to the moral lessons people might learn from them. He found the resemblance of coral zoophytes in the seas to terrestrial plants was too striking to be attributed to the only alternative of which the natural theologians could conceive, which was the operation of "blind, unguided, physical agencies acting by the appetencies of these organized beings."[12]

The likenesses between very different organisms brought Kirby to the subject of "parallel series" in animals. He was well aware that similarities in organisms were achieved in two different ways. Very often the resemblances between different species derived from their structural likeness as it does in two species of bees, for example. These resemblances were known as affinities. The alternative form of resemblance was called analogy since it was not due to structure but only to superficial resemblances between structurally quite different organisms. The resemblance between coral animals and vegetables was an analogous resemblance. These resemblances of analogy still retained a high importance for Kirby. Much of their significance came from their theological, lesson-teaching function.

> All these resemblances and approximations show, that the Great Creator embraced at one view all the forms to which he intended to give being, and created no individual without furnishing it with organs which give it some relation to others; or so moulding its outward form, as to cause it to represent some other to which it is clear it is not brought near by any characters common to both, that indicate affinity. What can more evidently and strongly manifest *design*, and that of a mind comprehending simultaneously the whole world of created beings, than thus to concatenate all link to link and wheel within wheel, through all their intricate revolutions and ramifications connecting and connected, and all the while reflecting others of a higher and lower grade with mimic features? This shows the hand, the art, the wisdom, the power, and the goodness of that unfathomable depth and immeasurable heighth of Deity, which comprehends all things and is comprehended by none; and to whom all things owe their being, and their form and their organs, and their several places and functions.[13]

In Kirby's orthodox opinion God was the Cause of Causes and the Designer of the Design. He was consequently, exempt from the human struggle to reconcile the parts of Creation with the whole. The naturalist sees an abundance of partial series, like the boughs and broken branches of a tree, which is a product of his human finitude.[14] Kirby believed that God had provided human beings with an indication of a path through the labyrinth of nature so that they could find out some of God's ways by searching properly. The "Ariadne's thread" that God has given is his system of laws, which the naturalist may follow. God has created according to a precise formulation all "link to link and wheel within wheel," rational laws attesting to a rational instigator of the laws.

Whenever Kirby or his associates discussed Law there was deliberate _double entendre_ in their usage of the word. For him Law conveyed both spiritual and natural connotations which helped the natural theologian to answer the difficult questions inherent in their position: Why does the Creator use the same ideas again and again? and Why does the Divine Mind resort to trickery and masquerade? The law of representation and the law of analogy were called upon to provide general solutions to particular cases of deceptive and repetitive resemblance.

The nineteenth-century natural theologians, of whom Kirby was among the chief, were masters of both empirical natural history and idealistic philosophy. That the world was full of strange and remarkable correspondences was visually verified; furthermore, the spiritual lessons from nature could be apprehended by the reasoning Christian believer. The striking similarity between the external form and color of many predators and their innocent herbivorous prey was a window for man to see into the reality of the demon kingdom.[15] The existence of two spirit domains, malevolent and beneficent, might be understood through the symbols of nature. The scorpion and venomous serpent were not themselves evil, but they stood as symbols of evil. People could, furthermore, learn from natural history the marvelous way in which the Creator wrought harmony out of the discord: species are appointed places and numbers which they are not permitted to surpass. In his discussion of predation, Kirby assured his readers that mercy rather than cruelty was the law of nature and that in the death of the herbivore to sustain the life of the carnivore not only were both species the beneficiaries, but also a lesson was taught on the prevalence of vicarious suffering even among insects. The symbols of Providential action were lavishly displayed in the facts of natural history. Even the insects, Kirby said, "Preach the Gospel."[16]

Analogy and Affinity

The same emphasis on the analogy between natural history and theology is also to be found in all the editions of the *Introduction to Entomology* written jointly with William Spence and first published in 1815.[17] The popular textbook was repeatedly reissued in new versions running through six editions by 1843. Much of Kirby's theory of analogies and disguises in nature was borrowed from the writings of William Sharp MacLeay.[18] Kirby had added facts and theological interpretations to MacLeay's system of classification. In his discussion of the history of entomology as a science Kirby went so far as to describe his own time as the "Era of MacLeay."[19] MacLeay devised his classificatory system on the basis of the two parallel forms of resemblance observed between animals: that is, those based on structure (affinities) and those superficial similarities not based on similar structures (analogies). Believing firmly that his was a natural system, MacLeay meshed the two forms of similarity into a single taxonomic system. It seemed to MacLeay that each natural group was invariably composed of five members and was circular in arrangement rather than dichotomously branched. MacLeay's system was generally referred to as the Quinarian System, and its advocates referred to as Quinarians. Its popularity was deservedly rather brief.

The system appealed to Kirby, and to several others, since it seemed to take the totality of the organism's relations and structure into consideration.[20] The similarities of analogy found a place in classification, not being automatically excluded. The observations that insects

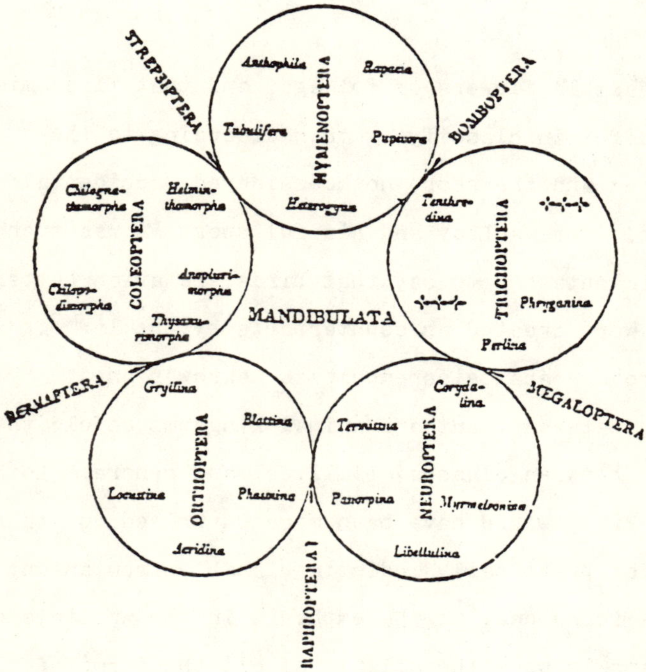

Fig. 1.--Classification of Mandibulate Insects according to the Quinarian System of W. S. MacLeay. (Reprinted from W. S. MacLeay's Horae Entomologicae, 1819, in William Kirby and William Spence, An Introduction to Entomology, Vol. 4, 5th ed., 1828, p. 479.

sometimes resemble flowers or foliage, and that flies mimic bees were useful in classifying them according to the Quinary System and therefore not considered "accidental" similarities. For MacLeay and his followers it was much more than a poetic metaphor to say that birds are avian butterflies; they were created as counterparts by the law of analogy. Broader analogies encompass narrower ones; resemblances between plant and animal kingdoms enfold the resemblances between classes, families, and genera. Both MacLeay and Kirby would have been more surprised not to have found a spider practically indistinguishable from an ant than to have found one.[21] The emphasis in the writings of the Quinarians is upon the utility of all the forms of similarity whereby one creature "mimics" another; they are bound together by the Final Cause of their being and both are used by animals in food-getting, or in defense, or in obtaining mates.

The Designer had implanted the analogous resemblances in organisms to serve a double purpose. They benefit the animal itself and provide the naturalist with materials to place it in proper relation to its allies. Means had been adapted to ends in these cases of Nature's Disguises as they were in all other sectors of natural philosophy. One of the most quoted examples of the matching of means to ends was the case of the parasitic fly, *Volucella*, which mimics

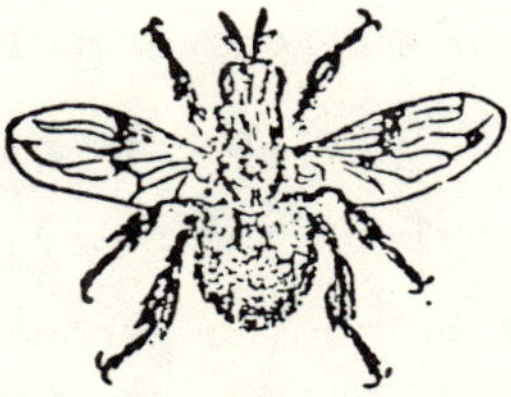

Fig. 2.--The fly, Volucella, which mimics the bumble-bee, a member of a different order. (From James Rennie, Natural History of Insects, Vol. 2, 1833, p. 147).

the bumble-bee.[22] Although this type of trickery might be deemed unworthy slyness on the part of the Designer by some naturalists, it was a favorite of Kirby's. He unabashedly asserted, "The Author of Nature provided that they may enter these nests and deposit their eggs undiscovered. ...Did these intruders venture themselves amongst the humble-bees in less kindred form, their lives would probably pay the forfeit of their presumption."[23] The parasite and the predator are the recipients of Divine provision as well as the prey, with a beneficial result to them all; justice is along the lines of the biblical principle wherein the "rain falls on the just and the unjust."[24] Parasites and carnivores function to keep their host species within a proper numerical balance and the weak are slain generally quickly and without pain. Animals which are not granted strength of structure, speed, or great fecundity are given strategem for their protection. The Volucella, therefore, is Deity's ne plus *ultra* for the humble-bee. The analogous resemblances in which nature abounds were clues for natural theologians to the nature of the Creator's provisions for the perpetuation of His creatures and, also, clues to the manner in which He had organized nature heirarchically.

The segregation of resemblances of analogy from those of affinity was not a simple task. Every naturalist who did taxonomy in the early nineteenth century knew that nature's disguises should spur him on to great efforts to avoid the errors which result from mistaking a resemblance of analogy for one of structural affinity.[25] The taxonomist who could not discern the difference between the two forms of visible likeness was almost certain to have a butterfly cabinet partly filled with unwanted moths. Commitment to natural theology did not make for a necessarily naive systematics. Relatively few describers of species were proponents of a strict numerical system like the Quinarian system. MacLeay's influence was most significant in calling attention to analogous resemblances and to their importance in animals' adaptations to the environment.

Mimicry and Analogy

Empirically the natural theologians could be as discriminating as the later Darwinians. It was clear to them in the 1830's that mimicries were superficial resemblances and that they function as adaptation for deception. One such entomologist, John Obadiah Westwood, knew nearly as much about the empirical aspects of mimetic phenomena as the Darwinians of the 1860's.[26] Westwood had studied the same group Bates surveyed in the Amazon, the family of models for mimicry, the Heliconiidae. He knew from his work as a taxonomist and as an artist that species of moths and of butterflies mimic the family Heliconiidae.[27] In his little volume, The Entomologist's Textbook, published in 1838, Westwood referred specifically to the resemblances between other insect species and the Heliconiids.[28] He called these mimicries "strange freaks of nature" which the unaccustomed eye might identify as the same species, but Westwood had seen through the deception and had catalogued the mimics with their structurally similar relatives even though they appeared superficially to belong to the family of models. Taxonomic error was not inherent in the natural theologians' position.

Like Kirby, Westwood accepted the premise that superficial resemblance indicated some deeper alliance, and, furthermore, he accepted the principle that similar ends had been wrought by the use of similar means. The analogous colors of insects might seem to be "strange freaks" to an uninitiated naturalist but they will be invariably found to be governed by harmonizing laws. The Designer has established that external colors shall be harmonized with habits; spiders which live on flowers, for example, almost always resemble them, whereas those which live on the ground are brown or gray. What an animal looks like is ruled by what it is supposed to do in the economy of nature. Philosophically the natural theologian held the opposite position from the later Darwinians; Westwood like other naturalist-theologians believed that the atheist holds that things make uses, but the Christian knows that it is the uses which make things.[29]

Westwood did not, of course, know everything that Bates did about mimicry. He was a taxonomist mostly limited to the museum, but nevertheless he knew that there were model species and that there were mimic species which were very dissimilar from the other members of their genera. It took someone with Bates's orientation toward biogeography to notice the precise overlap between the ranges of mimic and models and the gradual shifts of species along the Amazon River at the same degree of latitude. As a museum-based taxonomist, Westwood did not experience waning faith in the distinctness of species which Bates, Wallace, Darwin and other field naturalists experienced. Westwood retained his confidence in the reality of created distinct species until his death in 1893, and argued forcefully with both Wallace and Bates at the Entomological Society throughout the 1860's and '70's. Although Westwood was not a doctrinaire

Quinarian he was willing to use some of their arguments in opposition to the Darwinians during the 1860's.[30] The regularity with which analogous resemblances were observed and the way in which animals of different taxa represent each other functionally seemed to him to indicate that a general law was in operation. Like Kirby and other theologically inclined naturalists, Westwood's vision of nature was highlighted by the Unity of Design which was only further illuminated by nature's immeasurable diversity.

The _via media_ Westwood chose between total acceptance of analogy as useful in classification and its utter rejection as an object of scientific study was still too idealistic an approach to suit some naturalists. One such opponent of idealism in zoology, Hugh Strickland, believed that analogy should never be used to classify and that the repetitions of animal form were too erratic to be meaningful.[31] It seemed to him that systems based upon analogies were always disconfirmed by inductive investigation. No science could be done except _a posteriori_. Strickland urged observations as an antidote to the belief found among too many naturalists that nature was so much of a coherent unit that analogies of external appearance could be used to construct a model of species arrangement into which real species could be expected to fit _a priori_. He wrote in 1841, emphatically:

> We can no more speculate theoretically as to

> what groups are likely to remain undiscovered, than we can predict the discovery of rivers, lakes, or islands in any unexplored portion of the earth's surface. Both inquiries must be pursued in the same way, viz. by a careful induction of facts; and it will be found that there is much analogy between the process here recommended and that of a geographical survey.[32]

Concluding his article with a blast against MacLeay and his ornithological disciple, William Swainson, Hugh Strickland expressed his utter astonishment that naturalists had departed so far from the study of nature as she exists. With a poetic flourish Strickland likened nature to a wild, unpruned tree, luxuriant in its asymmetrical branchings; naturalists with their *a priori* systems had carved it into a clipped yew, a topiary, suited only to the formal Oriental garden.[33]

Although the use of analogous resemblances was challenged as having any appropriate role in taxonomy it was the subject of investigation and of a lively ongoing debate throughout the entire nineteenth century. There were those who based systems upon it, those who found compromises and those who abominated its use. In the process of argumentation, a large pool of data was produced and variously assimilated. The rejection of analogy as interpreted by William Kirby and the Quinarians was one phase of the pulling away from the natural theological way of looking at nature. Debates on the nature and significance of analogy have roots earlier in the nineteenth century, but became critical in the years just after the

Origin of Species; Bates's and Wallace's papers are part of that major shift in world-view. In the first half of the century, however, the subject of analogy as used by taxonomy was bound up tightly with Analogy as a general metaphysical concept from eighteenth-century works in natural religion. Interest in analogous resemblances of every kind had a long history before the abrupt and wrenching turn of events in 1859 or 1861.

Analogy and Natural Theology

The natural theology of the nineteenth century, no less than its medieval and ancient antecedents, was predicated upon the expectation of a general correspondence between God's (the Creator's) nature and qualities and the nature and qualities of his creatures. A relationship of analogy and general proportionality existed between the Contriver and the Contrived beings.[34] The argument based on analogy of proportion is necessarily imprecise and attempts to define it rigorously end in limiting the nature of the Designer. Theological writers were well aware of the difficulties inherent in natural theology, though they were seldom silenced by their doubts. Natural theologians tended to accept the analogy between Creator and nature as that between absolute and relative. Following the lead of Bishop Joseph Butler who wrote over a century earlier, George Field was willing to defend a position of relativism with respect to the evidences of Design in creation.[35] In the <u>Outlines of Analogical Philosophy</u>, it was admitted that although the existence of God is beyond question, the attributes of Goodness, Wisdom, and Power are only imperfectly seen in the natural creation. Field wrote in 1839:

> The regarding of things and thoughts as <u>absolute</u>, and overlooking their true <u>relations of analogy</u>, occasion the manifold errors and misunderstanding which mislead and distract both learned and unlearned.[36]

Like other Victorian natural theologians Field did not expect to find absolute perfection in the natural order which was relative in its very essence. Absolute perfection, Field held, in a relative thing, would be synonymous with nothingness just as absolute harmony in musical compositions is only accomplished by silence.[37] The natural theology of the nineteenth century was written by authors who were self-consciously assuming that they were the bearers of the burden of proof. If their arguments were inconclusive, it was because analogous reasoning is imprecise by nature and not susceptible to being made rigorous.

Joseph Butler, in the <u>Analogy of Religion</u>, 1736, stated that the argument from design depends upon analogical reasoning.[38] This admission was used against natural theology in David Hume's important book, <u>Dialogues Concerning Natural Religion</u>.[39] Hume demonstrated that the argument from design necessitates asking an infinitely regressing series of questions. Who designed the Designer? Hume's attack on natural theology indicated that every existing view of natural theology was philosophically flawed, but he was not able to prove conclusively that natural religion was impossible.

As a consistent sceptical philosopher, Hume did not believe what could not be positively asserted. Joseph Butler and his successor William Paley were of the opposite disposition. The absence of disproof was taken by them to amount to a positive indication that natural religion had been vindicated.[40] The followers of Butler and later of Paley accepted the possibility of natural theology as an assumption and were consistently prone to positive belief where disproof was not absolutely proved.

The practitioners of natural theology considered their work as empirical as that of any scientist. The Divine Craftsman had made intricate devices which could be researched and yielded information to the investigator both about the objects and their creator. Although nineteenth-century natural theologians were philosophically aware enough to realize that induction never yields conclusions of absolute certainty, they believed that the arguments based on design presented enough evidence to provide virtual certainty of God's reality and his participation in the world.

Natural theologians held that nature operates by rational laws which people can comprehend. The order of nature at every hand spoke loudly that a wielder of great power had turned chaos into creation.

The years in which the Bridgewater Treatises were written followed a period of rising reaction against the Deism of the Enlightenment. Orthodoxy was increasingly more conservative and evangelical in the years that followed both the Wesleyan revivals and the early stages of the evangelical renewal within the Anglican Churches.[41] The writers of the Bridgewater Treatisés are clearly Christians as distinguished from Deists. The eight volumes in the Bridgewater series were commissioned in the hope that the gathering of the empirical evidence of God's Goodness, Wisdom, and Power from the several quarters of the globe and of natural philosophy would prove conclusively that God is manifested in His work as well as in His Word. Each nominee accepted Paley's premises in his _Natural Theology_ of 1802 that nothing but a mind can account for the watch. Unless something as simple and mechanical as a watch can be explained on other grounds than by the action of intelligence surely nothing as complex as a living creature could be explained without recourse to intelligent purpose. The grandest analogy of proportion in the natural theological works of the 1830's was that between the products of Divine and human manufacture. They were reasoning by analogy and also reasoning on the reality of an Analogy between God and nature.

The three Bridgewater Treatises which discuss animals at the species and organism level of integration all show strong devotion to the cause of explicating the phenomena of the natural world with the tools of the theologian as well as with those of the natural philosopher. Unity of design and its corollary, the universal operation of natural laws, were the basic assumptions of Victorian scientists, who were both monotheists and Newtonians.[42] The single great principle that unites the Bridgewater authors on the subject of the natural and proper arrangement of species is the Law of Analogy. It is the principle which, if properly applied, makes it possible for men to retread the footsteps of the creator. Roget, Kirby and William Prout were each impressed by the combination of variety and unity in animal organization. The quotation from St. Paul's first letter to the Corinthians was placed on the flyleaf of volume two of Roget's Treatise on Animal and Vegetable Physiology with Reference to Natural Theology. "And there are diversities of operations, but it is the same God which worketh all in all."[43] The opposing "law of unity of composition" and the "law of gradation" created equilibrium in nature. The Designer's plan provided for the equal and opposite action of natural laws. Laws of variety in nature have been contained by a law of conformity to type like the one set out by Cuvier.

Although the use of analogy as a way of reasoning has no necessary connection to particular religious commitments, it was given a particularly prominent place in the writings of the nineteenth-century natural theologians who were devotees of an idealistic rather than an empiricistic philosophy of science.[44] The distinction between the use of analogy as a mode of ratiocination and as a term describing the relations between structures in two differing organisms is not always carefully drawn in nineteenth-century writings, since the use of analogy is accepted in both senses. Natural theologians who began arguments with the acceptance of the ultimate analogy between God and his contrivances continued their chains of reasoning by the "discovery" of structural similarities, or analogies, by which God has revealed the relations between animals. Analogy, with a capital A, was only the greatest in a chain of analogies; the analogies of nature point toward a greater cause as in idealism secondary causes point toward a qualitatively different and greater First or Final Cause. William Kirby, representative of the biological authors of the natural theological school, made the study of the individual analogies of nature a focal point for particular study. It should be noted that his discussions of natural similarities of forms and structures are not really separated from the indications they produce of the Final Cause, the producer of the Analogy. Natural history and worship of the Author are sincerely coupled.[45]

Natural history and natural theology are found together in Kirby's Textbook as well as in his Bridgewater Treatise. The Introduction to Entomology was highly influential in insect studies well into the third quarter of the nineteenth century. It was widely quoted and admired even after the premises which Kirby had fully accepted began to be questioned. The general confounding of philosophical distinctions in relation to the use of analogy as a term in science did not begin to be unraveled until Richard Owen's important work in 1843.[46] Owen attempted to use the terms with the precision of careful definition and with qualifying criteria. Both anatomy and taxonomy were beneficiaries of Owen's efforts as well as Strickland's, who wrote more directly in response to those who perpetuated the idealism and error traceable to W. S. MacLeay. With characteristic passion Strickland wrote:

> [Affinities] form an essential part of that magnificent plan of creation, which notwithstanding the amount of attention which Man has given to it may almost be said to be yet "to us invisible or dimly seen." Analogies, on the contrary, appear not to form any element whatever in the great system of Nature, but are merely examples of the recurrence of certain mechanical forms whenever the production of a certain mechanical action called for them; and so far from their being at or beyond the verge of human comprehension, we have seen that Man enjoys the high privilege of copying by these

> Analogies, at a humble distance the far transcendent works of his Maker.[47]

In the writings of both Strickland and Owen, the word "analogy" became a scientific term instead of an overarching principle. In Hugh Strickland's usage, it was the antonym of the term affinity, and in Owen's papers, analogy was the opposite of homology. It had in both cases the common element of serving to reveal a common function. As in the quotation above, Strickland did not assert that no teleological function at all was served by the presence of the functional, mechanical similarities of nature. However, both Strickland and Owen, as well as other morphologists and taxonomists recognized the need to separate the uses of analogy as a clue to the relations among animals and as an index of Divine Workmanship. The concept of analogy needed to be subdivided in order to be defined with any degree of precision. The first step was taken in the separation of structural and functional similarities. A second step was taken when the group of analogous resemblances was again pruned of a group of phenomena in which the superficial resemblance of analogy was without any apparent function. Strickland called these resemblances Iconisms, and some other authors used the word "Homomorphism" for a similar body of examples.[48] Included among the iconisms, instead of among nature's analogies, were the similarities of color and form of the Bee Orchis and the Bee, the Fungia and fungus, an ovulum and an egg.[49] Even with the efforts of Strickland, Owen and others, it was a very long time before the concept of analogy became anything like a rigorous, well-defined term.[50]

Many writers at midcentury were still philosophically the heirs of Paley and Bishop Joseph Butler. As the century proceeded there was a subtle, but a perceptible shift toward a less idealistic philosophy and toward a less exclusive emphasis on the created unity of nature. The adoption of empiricist philosophy and work on the discontinuity between animal groups was part of the movement in scientific circles which led, even before Darwin, to gradually loosened ties between natural history and the older style natural theology.

Catalogues of Disguises

In the 1850's the concept of analogy and analogous resemblances was treated at length in many scientific and popular articles and books. Each author used the very slippery concept in his own way; some were obviously teleological and idealistic, and some were empirical in their philosophical convictions. During the decade, the literature on the general subject of "Disguises in Nature" was accruing at a rate unabated by the realization that its exact meaning and use was in grave doubt. Joachim Schouw developed no less than five forms of these disguises and analogies in nature under the general heading, "Repetitions in Nature."[51] In 1852 he published his catalogue of analogies in a book written for a rather popular audience, The Earth, Plants, and Man.[52] Schouw had identified repetitions in nature based upon genetic affinity, geographic representation, historical continuity, taxonomic similarities, and the duplications of habits of animals. The repetitions of habits in otherwise different animals he denominated "mimicries."[53] Schouw was trying to deal with the apparent contradictions he had seen in nature. On the one hand, nature seemed like an overflowing cornucopia scarcely able to contain its abundance. On the other hand, nature seemed to be miserly with the ideas used in its productions. The repetitions of nature seemed to support the latter view just as the diversity of species supports the former. Schouw had no suggestion to offer as to what mechanism might explain how the duplications in the appearances of animals had been caused. He responded in the only way he could by stating that the laws of nature were transcendent and that words implying values in human life, like parsimony and profusion, were altogether inappropriate when applied to nature.[54] In Schouw's catalogue, the African Euphorbias mimics the American cacti since their similarity in form emerges from a fundamental similarity in habits of life. The comment on the remarkable similarity had to pass without a deeper analysis. In the 1850's there were people who were beginning to try to do better than merely to classify the analogies, or mimicries in color and form.

One such person was the Reverend H. H. Higgins, the vice-president of the Liverpool Literary and Philosophical Society, who made an attempt to decode the disguises of nature through an analysis of animal coloration patterns.[55] Higgins believed in 1857, as Kirby had much earlier, that the Author of Creation conceived of the animal and its habits as a single entity. Therfore, he contended, in the strict sense, no part taken by itself is an adaptation to any other; uses and structures are essentially inseparable. Except in cases in which an organism's colors were clearly designed to conceal the animal, it seemed to Higgins that the particular features of the color pattern had no particular relation to the survival of the species or to the place of the animal in nature's economy. He wondered, what did the Creator have in mind when he created the bands and spots of color on the wings of birds and butterflies? Higgins assumed that the color patterns were "determined by laws just as fixed and unalterable as those which regulate the construction of their most necessary organs."[56] After all, it was an article of faith that "Nature is never capricious." The rational laws which had been discovered to comprehend the universe strongly indicated that nothing was left without the control of a law. It did not seem probable that the wing patterns of insects had been the product of chance.

Higgins believed he had discovered the law on which the coloration of many animals might be explained. Most of the colors of animals could be resolved into two major categories: barred and spotted patterns. The banded colorational patterns seemed designed to convey the impression of strength in analogy to the strength of ribs. Spotted patterns, on the contrary, carry the connotation of lightness and lithe movement. The coded message of color had been utilized in some cases to accentuate the character of the animal, or, in other cases, to modify the impression by substituting the pattern of the opposite quality.[57] The more fragile, lighter antelopes, for example, tended to be barred or striped rather than spotted, so that they gave the illusion of strength which they would otherwise lack. The spots of the leopard, however, underscore its essential nature as a creature that embodies the qualities of speed and agility.

Higgins is not a major figure in the history of studies of animal coloration. His papers indicate, however, that a high level of interest in the ciphered messages of nature was present among naturalists before the _Origin of Species_ and Bates's publication in the Linnean Society's _Transactions_. Higgins's work also shows that the influence of natural theology was still strong when explanations for phenomena were offered and the use of the analogous resemblances of nature was still one accepted method of finding the unity of Design behind the bewildering variety.[58] Since patterns and colors were at once diverse and repeated, it seemed that this was one point in which the opposite tendencies in the Design were synthesized. Higgins wrote, "We see much of nature's painting, but are seldom permitted to examine her pencil."[59] Higgins's intuition was correct, that a study of butterfly wing patterns would yield many clues connecting the observable pattern and color with the causes of the particular designs.

Only two months before the Origin of Species appeared in print, a paper was given at the British Association in which the subject of the repetitions of the same forms and colors in animals was again broached. An entomologist and biogeographer, Andrew Murray, presented a paper: "On the Disguises of Nature: being an Inquiry into the Laws which Regulate the External Form and Colour in Plants and Animals."[60] Murray's paper indicates the point that knowledge of mimetic disguises had reached before evolution and natural selection became the major bones of contention whenever animal forms and colors were discussed. Murray, who had a prominent role in the opposition to Darwin and Wallace, held the belief in common with Reverend Higgins that the code of disguises could be elucidated. It was the regularity with which color and form were repeated that gave them both hope. Although Murray was well aware of the cases of disguises in nature that had been brought out by Kirby and Spence in the Introduction to Entomology, he justifiably believed that nobody had settled any of the questions raised by the similarities found among very different kinds of animals.[61] He was particularly interested in animals which respond to changes in their environmental circumstances by color changes. Murray held the opinion that the animals which vary seasonally, like the ptarmigan of the arctic, or animals which change color quickly, like the chameleon, were most likely to yield the secret to the mystery of animal disguises. The fact that disguises were sometimes variable seemed to indicate animal coloration was an expression of a "law of life."[62] A law of life is a dynamic law that permits the animal freedom to respond adaptively to its changing environment. It is not surprising that Murray wanted to be very sure that his law regulating animal disguises did not become confused with laws of species transformation like Lamarck's or Erasmus Darwin's.[63] Murray freely acknowledged that species vary within limits, but his law was not an evolutionary law.

Since the law governing nature's disguises had both general and special application, it must be an analogue of the Law of Gravitation, Murray reasoned.[64] Furthermore, since the disguises of nature are repetitions of shapes and colors where they are not anticipated on the basis of affinity, there must be a great Law of Attraction which evokes the aberrant appearance of color and form. The appearance of bee colors in the fly genus Volucella would then be explicable on the principle that there must be one cause behind one result. The same law, or a very similar one, produces analogous species whenever similar conditions of life are found. Murray gave the example of the similarly flightless birds of the grasslands and semi-arid districts of Africa, Australia, and South America.[65] The similarities among the ostriches, emu and rhea were striking. Murray knew from his work on zoogeography that analogous species were found in all the major regions of the earth wherever similar conditions of life were found. In many respects, the faunae of distinct continents were duplicates of each other, although they had no species in common. Murray was willing to go so far as to say that if conditions were absolutely identical, there was good ground for holding that the same species might be found in spacially isolated ranges.[66] The same law is in action throughout the world, everywhere effecting similar results. In his 1859 paper, Murray was taking the position that Creative Ideas are expressed in exclusive relationship to the external conditions into which the animal is introduced. Murray's law of attraction, determining the animal's external form and color, worked similarly to the action of magnetic attraction in which "Like begets like," and "Like draws to like."

> We have seen that in all the instances to which I have referred the external appearance of the animals bears definite relation to the appearance of the soil on which it lives, or the objects which surround it. It would appear as if there were a *genius loci*, whose subtle and pervading essence spread itself around, penetrating and impregnating the denizens of the place with its *facies*... How this mysterious influence may operate, it may be bootless to inquire; but is it unphilosophical, or inconsistent with the simplicity and grandeur of nature, to suppose that one great idea should contain the elements of the laws which regulate all the different constituent parts of the created world? That as by attraction the material particles forming the solid globe gravitate to each other--as by the action of chemical affinities a like law regulates their intermixture--so a law of attraction should operate on the immaterial world of thought and influence the phenomena of organic creation?[67]

Murray's style of asking and answering questions on the topic of analogous and protective resemblances between animals not related taxonomically bears much likeness to the work of his predecessors in both natural history and natural theology. He employed the analogy between unrelated species in the same way and he accepted the validity of the analogous method of reasoning. Although theological insights are not central to his paper, nevertheless, its author is clearly an idealist in contrast to an empiricist. Species are evoked in relationship to environmental criteria and are absolutely determined by a rigid comprehensive law which includes whatever apparent flexibility the animal has within it.[68] The chameleon is not free to turn whatever color it chooses any more than a stone can be thrown indefinitely along a horizontal line.

Murray had no mechanism to explain the collection of phenomena he had brought together under the general heading "the disguises of nature." The protective adaptations he had utilized were quite heterogeneous in character; most of them were employed for defense as general camouflage, but some, like the ostriches, were used to relate form to environment. The strand that held together the different kinds of "disguises" which nature uses was the law of attraction set up between animals and their circumstances. Murray's appeal to a Law rather than directly from data concerning organisms to conclusions involving the Creator puts him nearer the end of the natural theological spectrum represented by William Whewell than that held by William Kirby whose writings portray a more immanent and less transcendent picture of the Designer.[69] The argument from design held firm in 1859 and it was especially evident in the relations between organisms, and especially when different organisms seemed to be created according to a unifying pattern.

The topic of analogous representations and resemblances

came up again at the 1860 meeting of the British Association at Oxford, which is renowned for the encounter between Bishop Samuel Wilberforce and Thomas Henry Huxley. Cuthbert Collingwood read a paper to the BAAS bearing the title, "On Recurrent Animal Form, and its significance in Systematic Zoology."[70] It presented no very startling new data, although it does show that the subject had not been settled to general satisfaction. The lively debates concerning the nature of analogy, its role in taxonomy, and the revelations one can find of design or of law were still commanding attention and interest.

Bates's Mimicry without Design

These papers make it clear why Bates buried his discussion of the relationship between mimetic butterflies and the mechanism of natural selection in the middle and later pages of his paper. Bates knew that he must wield a two-edged sword. First, he was obliged to satisfy the taxonomists that he had not blundered on the relations of his butterfly species by mistaking affinity for analogy. That task must be carefully and convincingly done before he could accomplish his second task, which involved taking up the difficult question of the argument from Design in the context of animal colors and analogous resemblances. Bates knew that he had not stepped into a vacuum but into a maelstrom.

The answers to the old questions of the relation between Unity of Design and particular phenomena were met with the new Darwinian refusal to discuss final causes in the old terms. Bates's phenomena were really neither more nor less amazing than the many examples already known and explicated by earlier theologian-naturalists. The paper on the South American Heliconidae contained a flat refusal to acknowledge either a purposing Contriver or a predetermined purpose in the contrivance. Bates's study was nothing less momentous than a microcosmic refutation of the role of natural theology in natural history. The data used in Bates's paper was securely based on a long period of species identification and mapping. His data was fully and carefully presented, more from a geographical viewpoint than anyone but a traveller would present it, but otherwise much like the data that had been available for some years, as J. O. Westwood's textbook of 1838 indicates. The Darwinian debates were only very occasionally based upon new empirical data like a newly disinterred fossil <u>archeopteryx</u> which linked birds and reptiles. Most of the early arguments were about the reinterpretation of data. The proponents of natural theology's place in natural history had catalogued the Heliconiidae and its mimics appropriately, and it is not surprising that they had strong feelings about the meaning of the resemblances of analogy that illustrated the unity of nature and its Unifier.

The skilled attentions and determined efforts of several generations of eminent naturalists had been expended in the study of nature's disguises and analogous resemblances. In the light of long continued labor on the subject, the length and strength of the resistance to Bates's reinterpretation of these phenomena becomes more easily understood. The disguises of nature had become a kind of hallowed ground for natural theologians. These phenomena had been heralded as clearly manifesting the action of a Creator who makes similar animals flourish in appropriately similar conditions, but without repeating the same idea exactly. The author of the fifth Bridgewater Treatise, Peter Mark Roget, placed a quotation from Job on the flyleaf of his book as a rhetorical question having an obvious positive answer:

> Ask now the Beasts, and they shall teach thee, and the fowls of the air and they shall tell thee: ...Who knoweth not, in all these that the hand of the Lord hath wrought this?

In 1835 it would have been incredible to him, or to William Kirby, or to John Obadiah Westwood, or to the many others who labored in the joint cause of natural history and natural theology, that it would not be many years before naturalists in large numbers would answer that question, "Not I."

Notes--Chapter I

NATURAL THEOLOGY AND NATURE'S DISGUISES

1. Henry Walter Bates (born Leicester, England, in 1825; died London, England, 1892), "Contributions to an Insect Fauna of the Amazon Valley. Lepidoptera: Heliconiidae," Transactions of the Linnean Society of London 23:495-566, 1862. Bates, the son of a hosiery manufacturer in Leicester, was interested in natural history as a boy. In the company of his friend, Alfred Russel Wallace, Bates travelled to Brazil in 1818, where they collected together for about two years. Bates spent a total of eleven years collecting specimens along the Amazon River, embarking for England in February 1859. He was the author of a popular book of travels and natural history, The Naturalist on the River Amazon, published in 1863. From 1861 he served as assistant secretary to the Royal Geographical Society of London for twenty-eight years. Bates was elected president of the Entomological Society of London twice and Fellow of the Royal Society in 1881.

2. Bates, "Contributions...Lepidoptera: Heliconidae," p. 513.

3. Bates, "Contributions...Lepidoptera: Heliconidae," p. 514.

4. Two historians who discuss the relationship of evolutionary debates to the argument from design are Alvar Ellegard in Darwin and the General Reader (Gotenburg Studies in English VIII. Göteborg, 1958) and "The Darwinian Theory and the Argument from Design," Lychnos, 1956, pp. 173-192; and Walter B. Cannon, "The Bases of Darwin's Achievement: A Revaluation," Victorian Studies 5:109-134, 1961.

5. Asa Gray, "Natural Selection not Inconsistent with Natural Theology," Chapter III in Darwiniana (Cambridge, MA: Belknap Press, 1963), pp. 72-145. Originally published in Atlantic Monthly 6, July 1860, pp. 109-116; August 1860, pp. 229-239; and October 1860, pp. 406-425. Reprinted in 1861 with the title above.

6. Cannon, "The Bases...A revaluation," Victorian Studies 5:109-134, 1961.

7. Francis Henry Egerton, eighth Earl of Bridgewater, bequeathed in 1829 8000 pounds to the President of the Royal Society for the payment of author or authors to write treatises which would substantiate the truth of natural

religion from examples drawn from scientific studies. The President of the Royal Society, with the assistance of the Archbishop of Canterbury and the Bishop of London, nominated eight authors to write as many treatises. Each Treatise appeared under the prefatory notice that the testator had designated that treatises should be written: "On the Power, Wisdom, and Goodness of God, as manifested in the Creation; illustrating such work by all reasonable argument, as for instance the variety and formation of God's creatures in the animal, vegetable and mineral kingdoms; the effect of digestion, and thereby conversion; the construction of the hand of man, and an infinite variety of other arguments; as also by discoveries ancient and modern, in arts, sciences, and the whole extent of literature."

8. Peter Mark Roget, *Animal and Vegetable Physiology considered with Reference to Natural Theology*. 2 vols. (Philadelphia: Carey, Lea, and Blanchard, 1836). William Kirby, *On the Power, Wisdom, & Goodness of God as Manifested in the Creation of Animals & in their History, Habits and Instincts* (London: William Pickering, 1833). 2 vols. William Prout, *Chemistry, Meteorology, and the Function of Digestion Considered with Reference to Natural Theology* (Philadelphia: Carey, Lea, & Blanchard, 1836).

9. William Kirby (1759-1850) first president of the Entomological Society of London. Fellow of the Royal Society, and Fellow of the Linnean and Zoological Societies. The chief biography of Kirby was written after Kirby's death by John Freeman. *The Life of the Rev. William Kirby, M. A.* (London: Longman, Brown, Green, & Longmans, 1852).

10. Kirby's Bridgewater treatise contains long exegetical passages which are his attempts to reconcile the Hebrew uses of such words as "cherubim" and "heaven" with the operation of God with respect to natural order, pp. lxxvi-xcvi. Kirby's biographer asserts that the "suggestions on the sublime subjects" contained in the Introductory essay to the Bridgewater Treatise indicate that Kirby was a disciple of the mystical and somewhat cabalistic author John Hutchinson (1674-1737). Freeman, pp. 440-441.

11. Proverbs 25:2.

12. Kirby, *On the Power*... I, p. 191.

13. Kirby, *On the Power*... II, pp. 40-41.

14. Kirby, *On the Power*...II, p. 446.

15. Kirby made it clear that he did not believe that the poisonous serpents and scorpions were directly acted upon by demons, nor were harmless animals influenced by angels as some said. In explicating the reference by Christ in Luke 10:18-20 to the ability of his disciples to tread on poisonous creatures without being harmed, he clearly asserted that it was the spirits of evil of which the animals were figures to which Jesus referred. The idea that the animal which stings is the agent of demons forces the naturalist to hold the untenable belief, Kirby stated, that the bee is the agent of angels while gathering pollen and of evil spirits when it stings in response to danger. On the Power... II, pp. 232-234.

16. Kirby, On the Power... II, pp. 62-63.

17. William Kirby and William Spence, An Introduction to Entomology; or Elements of the Natural History of Insects; comprising an Account of Noxious and Useful Insects, of their Metamophoses, Food, Stratagem, Habitations, Societies, Motions, Noises, Hybernation, Instinct, etc. etc. 1st edition. London, 1815. 5th edition. 4 vols., London, 1828. 6th edition enlarged. 2 vols. London, 1843 (third after Kirby's death in 1856)

18. William Sharp MacLeay (1792-1865) was author of Horae Entomologicae: or Essays on the Anuulose Animals (London: 1819-21) Volume I in two parts. Several of Kirby's articles support MacLeay's philosophy of systematics. See Kirby's "A Description of some Insects which appear to exemplify Mr. William S. MacLeay's Doctrine of Affinity and Analogy," Transactions of the Linnean Society of London 1825, pp. 93-110.

19. Kirby and Spence, Introduction to Entomology IV p. 477. Although Kirby was powerfully influenced by MacLeay to accept his view of affinity and analogy, Kirby's classification work also shows the influence of the French naturalist, Pierre André Latreille (1762-1833). Kirby called Latreille's taxonomic system the "Eclectic." Introduction IV, p. 465. Carl Lindroth has pointed out Latreille's influence on Kirby in "Systematics Specializes between Fabricius and Darwin: 1800-1859" in History of Entomology, Ray Smith, Thomas Mittler, Carroll Smith, editors (Palo Alto, Ca.: Annual Reviews, 1973) pp. 119-154.

20. MacLeay's system was repeatedly discussed by a group which called itself the "Zoological Club." Made up of members of the Linnean Society chiefly under the influence

of N. B. Vigors, although Kirby was an active member from its inception in 1825. "Introductory Address, explanatory of the views of the Zoological Club, delivered at its foundation, November 29, 1823, by the Chairman, the Rev. William Kirby," Zoological Journal II, April 1825, pp. 1-8.

21. William Sharp MacLeay, "On Some New Forms of Arachnida," Annals of Natural History II, 1839, pp. 1-14; and A Letter on the Dying Struggle of the Dichotomous System (London: Printed by Richard Taylor, 1830).

22. The example of Volucella is mentioned by Kirby in the Introduction to Entomology II, Letter XXI (although the book was co-authored with W. Spence, Kirby was the author of the letter on the "Means by which insects defend themselves"). Freeman revealed the authorship of the chapters of the Introduction in Life of W. Kirby, pp. 308-309. Volucella is mentioned in On the Power... II, p. 326.

23. Kirby and Spence, Introduction to Entomology II, p. 442. 4th edition.

24. Matthew 5:45. Kirby is well aware of the struggle in nature and the checks to the natural tendency of species to increase beyond their food supplies. "Unless this tendency was met by some check, the world of animated beings would be perpetually encroaching upon each other, and would finally perish for want of sufficient food; and that the partial evils inflicted by one individual or one class upon another, to borrow a term from the Political Economist, proportions the demand to the supply." On the Power...I, pp. 143-44.

25. Resemblances of analogy were used in making taxonomic decisions at the species and sometimes at the genus level; only occasionally did they find use at the higher taxonomic levels. Darwin reflected the generally accepted way of working in his chapter on Classification in the Origin of Species. See pp. 427-428, where he wrote: "We can also understand the apparent paradox, that the very same characters are analogical when one class or order is compared to another, but give the true affinities when the members of the same class or order are compared one with another; thus the shape of the body and fin-like limbs are only analogical when whales are compared with fishes, being adaptations in both classes for swimming through the water; but the shape of the body and fin-like limbs serve as characters exhibiting true affinity between the several members of the whale family..."

26. John Obadiah Westwood (1805-1893), entomologist

and palaeographer. Founding member of the Entomological Society of London and Life President after 1883. First Hope Professor in invertebrate zoology at Oxford in 1861. Recipient of the Royal Society's Medal for 1855.

27. See plates I and V in vol. I, _The Genera of Diurnal Lepidoptera_ (London: Longman, Brown, Gray and Longmans, 1846-52) 2 vols. J. O. Westwood and Edward Doubleday. The mimics are properly catalogued with their relatives from whom they differ remarkably.

28. John Obadiah Westwood, _The Entomologist's Textbook: An Introduction to the Natural History, Structure, Physiology, and Classification of Insects, including the Crustacea and Arachnida_ (London: William S. Orr & Co., 1838) p. 304. Westwood gave the specific case of the _Acraea Thalia_ (a butterfly) and _Castnia acraeoides_ (a moth) which are a model-mimic pair seen flying together and _Castnia linus_ and _Heliconia psidii_, a mimic-model pair often confused by naturalists "whilst on the wing."

29. Westwood, _Ent. Text._ pp. 42-43. Westwood has paraphrased John Ray from _The Wisdom of God Manifested in the Works of Creation_ (London: S. Smith, 1704) 4th edition. The atheists are represented by Lucretius and French philosophers. Westwood concluded: "He would be but a sorry architect who, having completed the building of a splendid palace, had not, previously,to its erection, planned the uses of its various apartments, and adapted the size and situation of each to its intended uses." p. 43.

30. The debate between Westwood and Wallace and Bates is discussed in Chapter 4. Westwood attempted both to correct and to contribute to work on analogy and affinity, "Illustrations of the Relationships existing amongst Natural Objects, usually termed Affinity and Analogy, selected from the Class of Insects," _Transactions of the Linnean Society of London_,8: pp. 409-422, 1841; and "Observations upon the relationships existing amongst Natural Objects, resulting from more or less perfect resemblance, usually termed Affinity and Analogy," _Magazine of Natural History_, N.S., 1840, p. 141.

31. Hugh Strickland (1811-1853) was one of the founders of the Ray Society in 1843 and known mainly for the Code of Rules for Zoological Nomenclature which defined much of the taxonomic practice in the later 19th century. _Rules for Zoological Nomenclature Authorized by Section D of the British Association at Manchester, 1843_ (Edinburgh, 1863).

32. Strickland, "On the True Method of Discovering the Natural System in Zoology and Botany," _Annals and Magazine of Natural History_, _6_:184-194, 1841. Quotation from p. 189.

33. Throughout Strickland's writings the key word is "asymmetric." He does not conceive of how a symmetrical creation could cope with the immense irregularity of the conditions of existence; after all, he protested, "The design of creation was to form not a cabinet of curiosities, but a living world." p. 187.

34. Good though brief discussions of "Design Argument" and "Analogy" by Frederick Ferré may be found in the _Dictionary of Scientific Ideas_ and the _Encyclopedia of Philosophy_; as well as a chapter on the "Logic of Analogy" in _Language, Logic and God_ (New York: Harper & Row, 1969).

35. Bishop Joseph Butler (1692-1752) was bishop of Bristol and later of Durham. His contributions are discussed by R. W. Harris in Chapter 10, "Bishop Butler and Christian Orthodoxy" in _Reason and Nature in 18th-Century Thought_ (London: Barnes & Noble, 1969). See also Chapter V, Vol. I of Leslie Stephen's _History of English Thought in the 18th-Century_, 1876. (Reprinted by Harbinger Books, 1962). Chapter V, "Butler's 'Analogy'".

36. George Field.(1777?-1854), author of works on chromatography. _Outlines of Analogical Philosophy: being a primary view of the principles, relations and purposes of nature, science, and art._ 2 vols. (London: Charles Tilt, 1839) p. 469.

37. Field, _Outlines_...p. 474.

38. Joseph Butler, _The Analogy of Religion, Natural and Revealed, to the Constitution and Course of Nature._ (printed for James, John, and Paul Knapton at the Crown in Ludgate St., 1736).

39. David Hume,(1711-1776), _Dialogues Concerning Natural Religion_ (London, 1779).

40. William Paley (1743-1805), _Natural Theology: or, Evidences of the Existence and Attributes of the Deity, collected from the Appearances of Nature_ (London, 1802). Paley makes no mention of Hume's work. His argument is factual as opposed to philosophical and his view of the deity is a compromise between the view that God is immanent

and that he is transcendent. For Paley, God is immanent enough to perform miracles in the natural order, but transcendent enough to perform them according to preordained laws.

41. One article which discusses the conservative changes in England that impinged upon scientific endeavors is that by Norton Garfinkle, "Science and Religion in England, 1790-1800: The Critical Response to the Work of Erasmus Darwin," Journal of the History of Ideas 16:376-388, 1955. Walter Houghton in Victorian Frame of Mind (Yale Univ. Press, 1957), tended to underscore the importance of the identifica tion of Deism or atheism with the French Revolution.in the minds of Englishmen at the turn of the nineteenth century. The idea that God's existence could be documented empirically could hardly fail to appeal in the years following Paley's publication of Natural Theology in 1802.

42. For the Bridgewater Treatise writers at least the basic distinction between the problems of physics and of biology lay in addition of problems of ends and means to those of causes and effects that distinguish biological researches. Roget, for example, conceded that animals may be studied by reasoning from cause and effect as one would do in physics, when working with the mechanics of animal life. Organisms have valves in their veins, and muscles which work like pulleys as well as lenses which operate by the laws of optics. Whole animals must be understood in terms of the higher order of reasoning. Each species had been placed in a unique relationship in which the designer had adapted means to his own ends.

43. I Corinthians 12:6.

44. A good source is Alvar Ellegard's "The Darwinian Theory and Nineteenth-Century Philosophies of Science," Journal of the History of Ideas 18:362-393, 1957.

45. One example comes from Kirby's discussion of Functions and Instincts with respect to Man. "Since God created nothing in vain, we may rest assured that this system of representation was established with a particular view...In the first place, by observing that one object in nature represented another, he would be taught that all things are significant, as well as intended to act a certain part in the general drama; and further, as he proceeded to trace the analogies of character, in its two great branches, he would be led to the knowledge of the doctrine thus symbolically revealed--that in the invisible world there

are two classes of spirits--one benevolent and the other malevolent and mischievous; characters which, after his fall, he would find even exemplified in individuals of his own species." pp. 521-22.

46. Richard Owen, Lectures on the Comparative Anatomy and Physiology of the Invertebrate Animals (London: Longman, Brown, Green, Longmans, 1843).

47. Hugh Strickland, "On the Structural Relations of Organized Beings," Philosophical Magazine 28. 3rd series:354-364, 1845. Quotation from p. 362.

48. "Iconisms" was Strickland's term for accidental resemblances; he stated that it had been suggested to him by Rev. Dr. Ingram, President of Trinity College, and was preferable to the term "mimesis," which he had originally proposed. "On the Structural..." p. 356. Cuthbert Collingwood used the term "homomorphism." See "On Homomorphism; or, Organic Representative Form," Liverpool Literary and Philosophical Society, 1860, pp. 181-216.

49. Strickland's group would now be considered a heterogeneous grouping including accidental similarities as well as some mimetic cases. The bee orchis's resemblance to the bee serves to attract bees to aid in pollination of the flower. The fungia is an anthozoan, and ovulum is a genus of gastropod mollusc.

50. The subject received repeated treatment particularly by E. Ray Lankester in "On the Use of the term Homology in modern Zoology, and the distinction between Homogenetic and Homoplastic agreements," Annals and Magazine of Natural History 6:34-43, 1870; and by St. George Mivart, "On the Use of the Term "Homology," Ann. and Mag. of Nat. Hist. 6:113-121. For the 20th century continuation of the discussion see Alan Bovden, "Homology and Analogy: A century after the Definitions of 'Homologue' and 'Analogue' of Richard Owen," Quarterly Review of Biology 18:228-241, 1943; and "Homology and Analogy. A critical review of the meaning and implications of these concepts in biology," American Midland Naturalist 37:648-669, 1947. Also see Otto Haas and George G. Simpson, "Analysis of Some Phylogenetic Terms, with attempts at Redefinition," Proceedings of the American Philosophical Society, 90:319-347, 1946.

51. Joachim Schouw (1789-1852), Professor of Botany, University of Copenhagen.

52. Joachim F. Schouw, *The Earth Plants, and Man, Popular Pictures of Nature*, translated and edited by Arthur Henfrey (London: Henry G. Bohn, 1852) German ed. 1851.

53. Schouw, *The Earth*... p. 61. Schouw's "mimicries" were mostly examples which Kirby called "representation" and now are called examples of convergence.

54. Schouw, *The Earth*... p. 62.

55. Henry Hugh Higgins (1814-1893), author of *Notes by a Field Naturalist in the Tropics* (Liverpool, 1877). "On Colour Patterns in Natural Productions," *Proceedings of the Liverpool Literary and Philosophical Society* 1857, pp. 132-143.

56. Higgins, "On Colour Patterns..." p. 134.

57. Higgins, "On Colour Patterns..." p. 140.

58. Higgins shared with Kirby and other early writers the emphasis on the *message* that could be abstracted from animal coloration; he does not really confront the means or mechanism by which color patterns are developed in the animal morphologically.

59. Higgins, "On Colour Patterns..." p. 141.

60. Andrew Murray (1812-1878), naturalist, especially an entomologist. He is particularly known for his critical review of the *Origin of Species*, "On Mr. Darwin's Theory of the Origin of Species," *Proceedings of the Edinburgh Royal Society* 4:274-292, 1862; and for *The Geographical Distribution of Mammals* (London: Day & Son, 1866). "On the Disguises of Nature: being an Inquiry into the Laws which regulate External Form and Colour in Plants and Animals," *Edinburgh New Philosophical Journal* 11:66-90, 1860.

61. Murray chose Kirby's paradigm case of mimetic disguise, the resemblances between certain flies and the black and yellow bees, including *Volucella*. "On the Disguises..." pp. 68-70.

62. Murray, "On the Disguises..." pp. 79-80.

63. Murray, "On the Disguises..." p. 80.

64. Murray, "On the Disguises..." p. 77. "It is indeed very easy to say, that a special exercise of the

creative power has been used in these disguises in every separate instance: but if it is so, it is different from all we know of the other workings of Providence. A great law in all of them is set in motion, which, while regardless apparently of minor details and individual interests, yet, by its general working, most efficiently provides for them." p. 77.

65. Murray, "On the Disguises..." p. 82.

66. Murray, "On the Disguises..." p. 82.

67. Murray, "On the Disguises..." pp. 77-78.

68. By 1866, Murray was willing to admit that evolution had occurred in his Geographical Distribution of Mammals, Chapter 1, pp. 4-7. He employed another analogue from the physical sciences to explain the limits to variation: "I imagine that the law which secures this stability of species in INERTIA" (p. 7). Evolution could in some cases be accomplished in response to changes in the environment, without, however, the trial and error methods suggested in selection theories. Murray was one of Bates's and Wallace's major opponents on the subject of the evolution of mimetic resemblances. In letters to Nature he argued that hybridization and not natural selection produced the likenesses. See "Mimicry and Hybridization," Nature 3:154-56, 1870, and 186-187, 1871.

69. William Whewell (1794-1866), Astronomy and General Physics considered with Reference to Natural Theology (London: William Pickering, 1833). See also the evaluation of the image of the Deity as an Interfer in Walter Cannon's article, "The Problem of Miracles in the 1830's," Victorian Studies 4:4-32, 1960.

70. Cuthbert Collingwood (1826-1908), author of Rambles of a Naturalist on the Shores and Waters of the China Sea (London: J. Murray, 1868). "On Recurrent Animal Form, and its Significance in Systematic Zoology," Annals and Magazine of Natural History 6:81-91, 1860.

71. Job 12:7, 9.

Chapter II

BATES AND THE LABORATORY WHERE NATURE MAKES HER NEW SPECIES

Both of the discoverers of evolution by natural selection were field naturalists familiar with the subtleties of adaptation in wild organisms. Besides Charles Darwin and A. R. Wallace, several others who were early adherents to evolutionary points of view had been travellers and collectors. Thomas Henry Huxley had sailed on the HMS *Rattlesnake* for four years, and Joseph Dalton Hooker had travelled with Ross on an expedition to the Antarctic. Another less well-known naturalist, Henry Walter Bates, made a most important contribution to the Darwinian theory on the basis of eleven years' travel and research in the Amazonian region of Brazil.[1] The relationship between field experience, especially in the tropics, and the capacity to understand natural selection is particularly clear in Bates's writings.

Henry Walter Bates mapped species' distributions as he collected organisms for shipment to England. These studies of distributions of organisms in space and in respect to environments necessitated thinking in terms of continuity. He asked questions like these: at what point in a continuous area does variation reach the point of deserving its own species na[illegible]? What is a species? What is a variety? What is the meaning of the similarity between species in adjacent areas?

Work in taxonomy exercises the faculties of the mind that discriminate between animals instead of those that relate them. In systematics the worker must find criteria for a decision as to what is a species and what is only a variety. The analytic work of the taxonomist began to seem arbitrary to the synthetic biogeographers. Bates was like Wallace and Darwin in that he found it impossible to maintain faith in definite, unchanging, created species after having studied geographical distributions of animals.[2] In the work of all three naturalists, geographical relationships in space were related to historical, evolutionary relations in time. Lessons in biogeography are best learned in the tropics where diversity is richest and competition most intense, and they are only learned from wild species.

Natural selection cannot be appreciated at all unless a person has a prior understanding of the intensity and competition for food and reproduction. In confronting the burgeoning life of the tropical rain forest, Bates was ideally situated to understand struggle and differential survival. Bates's studies of Amazonian butterflies have had great significance in the debate concerning the reality of natural selection and its means of operation. Even today textbooks in introductory biology discuss Bates's discovery of mimicry by way of illustrating how natural selection produces modifications in species.

The history of Bates's publications on the mimicry in butterflies is almost exactly coextensive in time with Darwin and Wallace's publications on the theory of natural selection. Bates's first letter describing mimetic phenomena was printed in 1858, the same year the Darwin-Wallace papers were published by the Linnean Society of London.[3] The temporal similarity in the careers of mimicry theory and of natural selection theory is just one of the elements linking mimicry to Darwinism.

Geographer and Field Zoologist

In the latter months of 1857, Bates had been collecting specimens in the Upper Amazon region of Brazil. He was making a study of the geographical distributions of species near St. Paulo de Olivença (69.5° latitude) in comparison to those found in the vicinity of Fonte Boa (66.7° latitude), some two hundred miles downstream. In the process of collecting and comparing the faunae of the two village areas, he came upon a phenomena that both surprised and excited him; immediately he wrote to the *Zoologist* with a description of his findings:

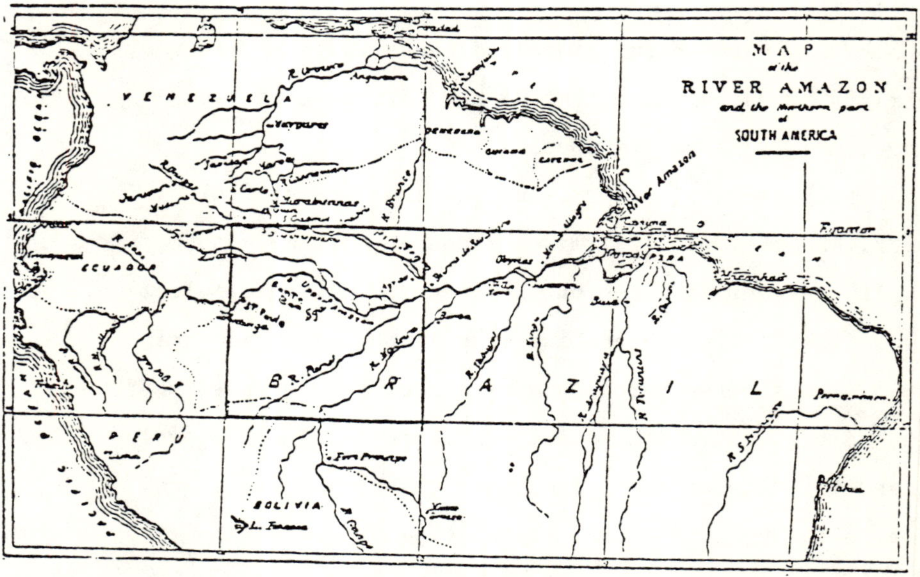

Fig. 3.--Map showing the location of Bates's collecting sites. (From A. R. Wallace, A Narrative of Travels, 2d ed., 1889, facing p. 1).

> Flying amongst the *Ithomiae* was now and then to be observed a *Leptalis*; I was very careful to secure every specimen, and the gathered series, now I come to examine them closely, have interested me as much as any other acquisition made during my excursion.[4]

Leptalis butterflies were members of the family Pieridae, which are relatives of the "garden whites," the European cabbage butterflies. These *Leptalis* butterflies were very different from their relatives; closely resembling instead the *Ithomiae*, a genus from a different family, the Heliconiidae.

It would have been startling enough to have found

species from different families that were similar in color and general form, but what Bates found was that the *Leptalis* and *Ithomiae* which looked alike lived in identical ranges. He could find neither species as far as one hundred-fifty miles downstream. Bates wrote: "It would seem then almost correct to say that at Ega and other stations these new *Leptalis* are not found because the *Ithomiae* to which they correspond are also absent."[5] Here, then, was something new--species of two different families which could not be distinguished in flight. Bates admitted that he could "scarcely restrain an exclamation of surprise" upon opening a net containing what he thought must be an *Ithomia*, but "found when in the net, to be a *Leptalis* mimicking it."[6]

Indicating that he had given thought to cases of resemblance between even more distantly related animals than between families of butterflies, Bates described the resemblance between different superfamilies, a Bombycide moth and a Heliconiide butterfly. Bates concluded the letter to *The Zoologist* with a characteristic understatement, "These analogies to me appear one of the most beautiful phenomena in nature."[7] Without feigning any hypothesis about why the two structurally dissimilar insects were so much alike in color, he abruptly terminated his discussion of the subject with an apology. "I am afraid," Bates demurred, "I am occupying too much space with the details of my doings in the Diurnes."[8]

Bates published nothing further on the "remarkable analogies" until after his return to London in 1859. His return to England was timed perfectly, although not purposely, for him to take part in the earliest debates about species and their origin following the publication of the *Origin of Species*. The active phase of field collection and biogeography ended for Bates upon his return to England, but the data he collected there provided him source material for many papers. The specimens of mimetic butterflies he had shipped back to London gave him the information he needed for his insightful statements about the evolution of insect colors and the relationship it had to the origin of species. The phenomenon of mimicry was seen by Bates in the Amazons but it was in the relative leisure of England and during his period of recovering his health that Bates wrote the paper explaining mimicry in Darwinian terms.

Contributions to an Insect Fauna of the Amazon Valley

The first task Bates had set himself was the publication of a series of papers summarizing his eleven years abroad. The papers were to have the general title "Contributions to an Insect Fauna of the Amazon Valley" with varying subtitles. The first was "Part 1. Diurnal Lepidoptera" and appeared in the *Transactions of the*

Entomological Society of London.[9] In this publication Bates took a single group to study in depth in order to learn lessons from it that might apply to other groups. The group chosen was the large genus of butterflies, *Papilio*.

He had learned that the Amazonian species of *Papilio* were more closely related to the species from the north in Guiana rather than to those from the southern parts of Brazil. This finding was important because it opposed Darwin's assertion in the *Origin of Species* that even the equatorial region was chilled during the glacial epoch.[10] Bates held to the contrary belief that the large number of endemic species of *Papilio* indicated that no significant climatic change had occurred in a much longer time than since the last glaciation. Any significant climatic change would have caused the extinction of the endemic tropical forms which are still present in tropical America.

Although the first "Contributions" paper contained a specific argument against Darwin, it contained other points that clearly supported Darwin's view of species and the changes they may undergo. Their difference on the refrigeration of the tropics was minor; the similarity of their views on the nature of species was far more important. Bates wanted his readers to understand at the outset what his principles were in making the designations of "species" and "variety." He wrote:

> I have been constrained to adopt the principle of treating every distinct and constant local variety as a separate form; giving it a separate name, but pointing out always what I thought to be its true relation to the allied species.[11]

Giving each well-marked, permanent variety a separate name was the only feasible course to take despite the effect the different names had in obscuring the very great similarities between two species or between species and its varieties.

It was immensely difficult to make clear-cut distinctions between varieties and species. The problem was multiplied by the fact that different species, even in the same genus, differed in their susceptibility to variation. Nothing, it seemed, remained constant in nature. Bates described his frustration with respect to the butterfly, *Papilio Bolivar*.

> Susceptibility of change, power of adaptation with or without change to new local circumstances, are qualities or characteristics of species, just as much as bodily structure or peculiar instincts...After becoming thoroughly familiar with the productions of one region during several years residence, I have at different times, removed to another several hundred miles distant, and have been then obliged to notice the changes of appearance that many of my old friends had put on; some more, others less; whilst some had assumed quite the form of new and distinct species.[12]

Nature abounded in diverse forms of deception that were concentrated in the genus, *Papilio*. The *Papilio* butterflies manifested a wide spectrum of variations and demonstrated the process of gradual geographical speciation.

Bates could not have been more clear that species arise from varieties. He had acknowledged that the distinction between species and varieties had been made for reasons of convenience, since he needed to have a name for each well-marked variety in order to make comparisons between faunal lists of different places. Bates carefully avoided the common error of ascribing to nature distinctions made for the convenience of the naturalist, writing simply:

> There is no absolute, well defined distinction between these different grades of relationship, and therefore there are in the following list many forms which I consider merely a matter of individual opinion under what category to treat them; some varieties might be considered mere individual differences, some local varieties as true or complete species, and so forth.[13]

Bates was more than slightly overwhelmed by the extraordinary diversity of the tropical genus. Darwin's image of the "tangled bank" was a poetic image to leave in the mind of readers at the end of the *Origin of Species*, but for Bates the tangle was source of true frustration.[14] The taxonomist reduced the snarl of diversity to a list of names. Bates stated in his last paragraph, "Nature as far as species are concerned, wears a false air of simplicity in all our lists and monographs."[15] Bates had learned his lesson from nature that species' names are superimposed on nature's productions and are not in any way a rediscovery of a plan inherent in nature.

The years in the Amazon had destroyed any confidence in absolute species Bates might have had. Not only was the line between variety and species hard to draw, but there were resemblances between unrelated forms and forms differing remarkably from members of their own genus, as in the species of *Papilio pausanius* and *zagreus*.[16] The studies in geographical distribution and variation Bates had done with *Papilio* were tantamount to descriptions of geographical speciation. He did not make the implications of his research for Darwin's theory explicit in the 1860 paper even though they were quite clear.

The connection between geographical distribution and geographical speciation so notable in Bates's writings even before 1859 is also present in Wallace's papers. From the beginning of his travels as a naturalist, Wallace had held that species arise in the same way that varieties do. He had made that clear on several occasions before the publication announcing his view of natural selection in 1858.[17] Both Bates and Wallace were working from similar data and with common assumptions, particularly on the identity of well-marked races and species. Wallace's letter to Bates on January 4, 1858, indicates the views he held in common with Bates:

> He [Darwin] is now preparing for publication his great work on species and varieties, for which he has been collecting information twenty years. He may save me the trouble of writing the second part of my hypothesis by proving that there is no difference in nature between the origin of species and varieties, or he may give me trouble by arriving at another conclusion but in all events his facts will be given for me to work upon. Your collections and my own will furnish most valuable material to illustrate and prove the universal applicability of the hypothesis. The connection between the succession of affinities and the geographical distribution of a group, worked out species by species, has never been shown a as we shall be able to show it.[18]

If they could confirm the gradual establishment of a distinct geographical race of a species both Bates and Wallace felt that they would have demonstrated evolution.

Each man had his own style of accomplishing the common goal which reflected the differences in their personalities. Bates was the more inductive in his approach. He required larger bodies of data before drawing theoretical conclusions. Bates was more a confirmer of theories than an initiator of them. Bates would have been extremely unlikely to send off a little paper for publication containing a controversial new theory with only a few details and supporting facts as Wallace had done with his paper of 1855, "On the Law which has regulated the Introduction of New Species."

The papers which Bates authored show his tacit assent to evolutionary principles. The letters written to Wallace following the publication of "On the Law" illustrate the differences in style and approach to the evolutionary zoology they were both doing. It is clear in the letters that the two had discussed evolution during their joint travels in the Amazon, or perhaps even earlier. Bates wrote, "The theory I quite assent to as you know, was conceived by me also, but I profess that I could not have propounded it with so much force and completeness."[19] The presentation of data with a few cautious generalizations was much more in keeping with Bates's sense of what was needed in science and what was appropriate in a scientific paper. The letter to Wallace continues:

> Many details I could supply, in fact a great deal remains to be done to illustrate and confirm the theory: a new method of investigating and propounding zoology and botany inductively is necessitated, and new libraries will have to be written; in part of this task I hope to be a labourer for many happy and profitable years. What a noble subject would be that of a monograph of a group of beings peculiar to one region but offering different species in each province of it--tracing the laws which connect together the modifications of forms and colour with the _local_ circumstances of a province or station--_tracing_ as far as possible the actual _affiliation_ of the species.
>
> Two such groups occur to me at once, in entomology, in Heliconiidae and Erotylidae of South America.[20]

Even in 1856, Bates was perspicaciously centering his attention on groups of animals which would shed light on the process of evolutionary change. His most important contribution to entomology and to evolutionary zoology came out of the study of the Heliconiidae. Not only would they yield useful data on geographical speciation and local adaptation, but these were the models for what came to be called "Batesian mimicry."

Mimicry and the Darwinian Theory

Bates published his data on the Heliconiidae as part of his series of "Contributions to an Insect Fauna of the Amazon Valley. Lepidoptera: Heliconidae."[21] It was published by the Linnean Society in the Proceedings for 1862, where it was adorned by two beautiful plates. The paper began with a review of the classification of the Heliconiidae and included suggestions for its revision setting out the characteristics that seemed to Bates to be diagnostic of the species and consistent with the geographical distributions of species. Any reader who failed to penetrate deeper into the paper than the first six pages would fail to perceive that it contained far more than was promised in its modest title.

After his exposition of Heliconiide taxonomy, Bates explained his findings on the geographical distributions of species, repeating his belief that the field zoologist has insights into nature not available to the taxonomists in museums. The taxonomy of the Heliconiidae, as well as that of other groups, was being bungled by "closet naturalists" who could not appreciate the variability of species since all they saw were preserved specimens manifesting what looked like discontinuous variations.

> It is only by the study of variable species that we can obtain a clue to the explanation of the rest. But such species must be studied in nature, and with strict reference to the geographical relations of their varieties. Many closet naturalists who receive disconnectedly the different varieties in any group, treat them all as independent species; by such a proceeding, it is no wonder that they have faith in the absolute distinctness and immutibility of species.[22]

The zoogeographical picture of the Heliconiides indicated that the family were the models for many mimetic resemblances by species very distantly related to them. The *Leptalis* genus from the family Pieridae had many species mimicking the Heliconiidae. Bates had received corroboration of his opinion from Wallace, who had already observed mimicry in the Old World Tropics.[23] Wallace had found that the family Danaidae, close relatives of the Heliconiidae, were the models for mimics in the Malay Archipelago.

Bates had concluded that the Heliconiidae were the models for mimicry since they retained the coloration characterizing their own group, whereas the *Leptalis* species had widely departed from the color and form typical of their closest relatives and had taken on the coloration found in Heliconiidae. Mimetic resemblances, then, were resemblances of "analogy," because they had not resulted from common ancestry. Analogous resemblances, like mimicry, are superficial likenesses only, not resemblances resulting from structural relationships. Bates explained the mimetic analogies by the use of clear and original metaphor, "An idea of what is meant may be formed by supposing a Pigeon to exist with the general figure and plumage of a Hawk."[24]

The colored plates demonstrated that a rather large number of species were participants in mimicry. Furthermore, Bates showed that the ranges of the mimics were limited by the ranges of the models. The color and form of the *Leptalis* mimics followed the changes in Heliconiid models. It seemed to him that the speciation in mimic species was contingent upon the variation in models. Referring his readers again to the plates, Bates showed that some mimics are better than others in duplicating exactly the colors of models.[25] (See Plate II.) The inference from that observation, for an evolutionist, is that mimicry is acquired gradually. Mimicry is not conferred on a species but involves a gradual acquisition of adaptive variations.

Bates still had not explained what might motivate a change in a species from its own typical coloration to that of a different family. To explain what he called the "meaning or final cause of these analogies" Bates reminded his readers of the cases of mimicry known to every entomologist and amateur collector.[26] All of them were familiar with the

PLATE II

MIMETIC ANALOGIES BETWEEN SPECIES
OF LEPTALIS AND HELICONIIDAE

(From H. W. Bates, "Contributions to an Insect Fauna of the Amazon Valley. Lepidoptera: Heliconidae," *Trans. Linn. Soc.* 23, Table 56).

EXPLANATION OF PLATE II

Figures 1, 2 and 3 are geographical varieties of *Leptalis theonoë* which vary in the degree of their imitation of the model, *Ithomia orolina* (Fig. 3a).

Figure 4 depicts an edible *Leptalis theonoë* variety resembling *Ithomia ilerdina* (Fig. 4a) which is unpalatable to many insectivorous animals.

Figure 5, *Leptalis nehemia*, shows the normal form of the family Pieridae (called "Garden Whites" in England), to which the *Leptalis* species belong.

Figures 6, 7 and 8 represent three species of *Leptalis* (*L. theonoë*, *L. amphione*, and *L. orise*) which have models in three different unpalatable genera of Heliconiidae; *Ithomia virginia* (Fig. 6a), *Mechanitis polymnia* (Fig. 7a), and *Methona psidii* (Fig. 8a).

Introduction to Entomology of Kirby and Spence, in which the resemblance of the Hawk-moth to the Bee is described.[27] Also, Dr. Cuthbert Collingwood had written a more recent article describing these mimetic resemblances and called them "recurrent animal forms."[28] Any resemblance of an animal to something else must be explained in terms of adaptation. The resemblance of a caterpillar to a dry twig or the resemblance of the Hawk-moth to the bee were protective adaptations. If a moth resembles a wasp, it is logical to suppose that the imitation serves to protect the moth by deceiving moth predators into avoiding the defenseless mimic. Bates asked rhetorically, "May not the Heliconide dress serve the same purpose to the _Leptalis_? Is it not probable, seeing the excessive abundance of the one species and the fewness of the other, that the Heliconide is free from the persecution to which the _Leptalis_ is subjected?"[29] There could be no other answer than "yes." Besides moths resembling bees and wasps, Bates had seen caterpillars with both behavioral adaptations and coloration like a small snake. All of these deceptions were similar in their protective function.

The similarities of Bates's explanations to William Kirby's are startling. The following statement by Bates will serve to illustrate this:

> Every species in nature my be looked upon as maintaining its existence by virtue of some endowment enabling it to withstand the host of

> adverse circumstances by which it is surrounded. The means are of endless diversity. Some are provided with special organs of offence, others have passive means of holding their own in the battle of life. Great fecundity is generally of avail, added to capabilities, active or passive, of wide dispersion; so that when the species is extirpated in one part of its area of distribution, the place is refilled by migration of individuals from another part. A great number have means of concealment from their enemies, of one sort or another. Many are enabled to escape extermination, or obtain subsistence, by disguises of various kinds; amongst these must be reckoned the adaptive resemblance of an otherwise defenceless species to one whose flourishing race shows that it enjoys peculiar advantages.[30]

If Bates had ascribed the adaptations to an all-wise Providence, he would have been paraphrasing Kirby, Spence, or Paley; but he did not. Bates found an explanation of species that did not require an appeal to teleological explanations nor had he any need to posit the action of occult or divine forces. Individual variation and the isolation of local races provided the only prerequisites for evolutionary changes in species. The development of mimetic disguises was Bates's chief example of adaptive change of organisms which indicated to him how general evolution occurs. Mimicry studies illuminated the motivating reasons behind species change. The advantage for an edible insect, like <u>Leptalis</u>, in resembling the Heliconiidae which were rarely eaten by insectivorous animals, was easy to understand. The same advantage was found in <u>Leptalis</u> butterflies resembling the Heliconiidae as the moth obtained in resembling a stinging wasp. Mimicry had a utilitarian function, and it was possible to achieve on the basis of individual variation and differential survival of individuals (natural selection).

Bates was able to assert on the basis of his own experience that models among the Heliconiidae were rarely persecuted by the usual predators of butterflies. He had not seen them taken by birds, dragon-flies, lizards, or predaceous flies, even though they were slow fliers. Wallace had been consulted for any similar findings in the Malay Archipelago. Bates placed a footnote in his paper that both the Danaoid Heliconiidae of South America and their analogues in the Malay Archipelago, the _Euploeae_, possessed a peculiar, slightly unpleasant, odor, by human perception.[31] Bates needed to demonstrate conclusively that the Heliconiidae were protected from their enemies by some substance in their bodies of which their coloration was a signal to potential predators.[32] Besides the odd smell, which even a human being could perceive, Bates found confirming evidence in the fact that the models of mimicry were much less subject to the ravages of vermin in entomological collections than most other specimens were.

Being the careful inductive reasoner that he was, Bates established first that species change, second that they have adaptive reasons for changing, as in the case of the mimic to escape extermination by predators. He had to explain next how that change had occurred.

Bates began forthrightly, "The explanation of this seems to be quite clear on the theory of natural selection as recently expounded by Mr. Darwin in the 'Origin of Species.'"[33] The direct effect of the environment could not explain the species changes, since the land area which had been investigated was not broken up by barriers, its vegetation and climate were similar throughout. The sudden generation of species by sports or large mutation also could not explain the gradual sequences of changes in species that Bates recorded. Mr. Darwin's were the only possible principles adequate to explain the phenomena: the alternative had been eliminated.

> It is clear, therefore, that some other active principle must be here at work to draw out, as it were, steadily in certain directions the suitable variations which arise, generation after generation, until forms have resulted which, like our races of *Leptalis Theonoë*, are considerably different from their parent as well as their sister forms. This principle can be no other than natural selection, the selecting agents being insectivorous animals, which gradually destroy those sports or varieties that are not sufficiently like *Ithomiae* to deceive them.[34]

The predators who were the agents of natural selection were constant in their pressure on the edible species. Any individual not presenting an adequate counterfeit would be taken. Any variations which improved the likeness to the unpalatable model would be spared longer, giving it greater opportunity to leave offstpring. Bates described the process of natural selection as being like that of the horticulturalist who desired to create new colors of a particular flower.[35] He can produce the new colors by "rogueing" the flower beds of unwanted variants. Plants with flowers showing colors in the direction of the desired colors are allowed to reproduce. The process must be repeated as long as necessary to eliminate unwanted colors and to perfect the desired tints. In the production of mimetic analogues, the gardener is the dragon-fly, the lizard, or the bird.

Nature was irrepressibly spontaneous in providing variations on which the predators worked. There was no need to postulate a Cosmic Gardener. Natural variation was subject only to natural selection, not to predestination. Bates concluded his paper with the statement that the deceptive mimicries had been seen by other naturalists and ascribed to innate principles.

> On examination, however, these explanations are found to be untenable, and appearances which suggest them illusory. Those who earnestly desire a rational explanation, must, I think, arrive at the conclusion that these apparently miraculous, but always beautiful and wonderful, mimetic resemblances, and therefore probably every other kind of adaptation in beings, are brought about by agencies similar to those we have here discussed.[36]

The importance of his data and conclusions were clear to Darwin, and he was justifiably delighted.

At last there was a new body of data that specifically supported natural selection. Darwin conveyed his enthusiasm for Bates's article in superlatives.[37] He wrote to Bates, "In my opinion it is one of the most remarkable papers I have ever read in my life. The mimetic cases are truly marvelous, and you connect excellently a host of analogous facts."[38] Darwin only regretted that Bates had entitled his paper "Contributions to an Insect Fauna" instead of calling attention to the mimetic resemblances the paper described. But it was a gentle reproof. Darwin wrote from his personal experience, "Your paper is too good to be appreciated by the mob of naturalists without souls, but rely on it, that it will have lasting value."[39]

Surely Bates was not surprised that Darwin wished he had chosen a different title. His letter of reply to Darwin does not discuss the chiding remark, but encourages Darwin to advance additional criticisms and thanks him for his generous reception of the treatise.[40]

Bates's simple title and Darwin's response to it underscore the differences between the goals and methods the two men employed. Bates did not perceive that he was doing the same kind of work that Darwin was doing. He was doing just what he promised to do--contributing to a fauna of the Amazon Valley. The faunal description Bates was writing differed in character from most that preceded it in that it was not only a list or even only a geographical map. Bates started at the roots of zoogeography with criticisms of taxonomy and the criteria utilized for making species in the light of their distributions. The mimetic resemblances he had found were also placed in their proper context in his paper. They were, first of all, dramatic cases of range limitation; mimics were limited in their range by biological barriers: the distributions of their models. The limitations of the geographical ranges of mimics were enforced by the operation of natural selection, since the mimic would have no protection outside the area in which predators recognized the colors of unpalatable butterflies.

Bates had not taken up his subject out of pro-Darwinian polemical motives. He was not a person who enjoyed heated debate in the way that Huxley did. Bates would never have qualified as a "bulldog" for Darwin. His addresses show that he much preferred to play the role of the reconciler of factions. Bates believed that truth would ultimately prevail and that the gathering of data would provide even the most hesitant entomologist with incontrovertible evidence indicating that evolution occurs by means of natural selection.

Darwin knew that the _Proceedings of the Linnean Society_ would be read by many of the professional biologists, but he felt that its message ought to be repeated in another publication. He published an anonymous review of the

"Contributions" in the Natural History Review. Darwin emphasized the bearing of Bates's findings on the creationists' position.

> Hence the creationist will have to admit that some of these forms have become imitators, by means of the laws of variation, whilst others he must look at as separately created under their present guise; he will further have to admit that some have been created in imitation of forms not themselves created as we now see them but due to the laws of variation! Prof. Agassiz, indeed, would think nothing of this difficulty; for he believes that not only each species and each variety, but that groups of individuals, though identically the same, when inhabiting distinct countries, have been all separately created in due proportional numbers to the wants of each land. Not many naturalists will be content thus to believe that varieties and individuals have been turned out all ready made, almost as a manufacturer turns out toys according to the temporary demand of the market.[41]

Bates's strong supporting data inspired Darwin to use a more derisive tone for his opponents than he usually employed in his writings. The evidence of geographical speciation and of the natural selection of adaptive variations made Darwin hopeful that the data would persuade. He exclaimed:

> It is hardly an exaggeration to say, that whilst reading and reflecting on the various facts given in this Memoir, we fell to be as near witnesses, as we can ever hope to be, of the creation of a new species on this earth.[42]

Asa Gray also appreciated the importance of Bates's study of mimicry. He found it ideal for the sport of "Agassiz-baiting." Like Darwin, Gray could see that Bates's study was in full opposition to Agassiz's notion of "centers of creation." Gray wrote to Darwin, "The copy of Bates' paper addressed to me by the author has come."

> It is fascinating: He at least is a full believer in Nat. Selection & the illustrations he brings to your aid are the prettiest possible. I keep it in pickle along with Decandolle on oaks & some other matters to make the subject of the evening for our Scientific Society which meets with me a fortnight hence, when I hope to torment Agassiz delightfully, and then I will myself draw up an abstract of the interesting matter for Silliman...[43]

The copy of Bates's paper addressed to Gray "with the author's kind regards" is in the Museum of Comparative Zoology library at Harvard. In it, Gray marked the passages he wished to appear in the extract, omitting the first four and one half pages, which contained Bates's assessment of the taxonomy and zoogeography of the Heliconiidae. The passages Gray excerpted to use in the polemics with Agassiz were the ones in which Bates makes his declarations of belief in the theory of natural selection. In the hands of other pro-Darwinian biologists, Bates's paper takes on a more theoretical character than it had in its original publication. The taxonomy and geography were central features in the "Contributions." Dedication to data and inductive science did not always characterize other Darwinians to the degree that it characterized Henry Walter Bates.

The Round-Robin Correspondence of Bates, Hooker, and Darwin

Bates's "Contributions" paper brought him into contact with several important naturalists who were early believers in the Darwinian theory of evolution. Among them was Joseph Dalton Hooker, who wrote to Bates expressing his great interest in the recent paper, and directed a friendly criticism to him.[44] Hooker believed that Bates was somewhat confused on the distinction between the influence of the environment on organisms and the effect of natural selection. Darwin, who was already a long-term correspondent of Hooker's, had begun to direct questions to Bates on a variety of subjects. A series of letters circulated among the three correspondents in a round-robin fashion.[45]

The letters present an interesting view of three evolutionists applying a theory to which each person was committed to different fields of biological investigation. Bates was about to demonstrate his mimetic butterflies before the Linnean Society; Hooker was working out his theory of centrifugal variation; and Darwin was grappling with use and disuse for his study of variation in domestic species. One of the most intriguing aspects of the exchange is the central role Joseph Hooker played as explicator and defender of natural selection to Darwin and Bates.

Bates was trying to prepare for objections to the theory of the evolution of mimicry by natural selection. He wanted to review with Hooker the best defenses of the natural selection position. The precise nature of the mechanism of natural selection and its positive role in the making of new species, in contrast to its negative role in weeding out the unfit, were still not fully clear to him. Bates's letter reads:

> It appears to me that the whole question of natural selection might be put to the test by a close examination of these few insects. They show that the same species varies in a different way in different localities, therefore that natural selection does not explain everything, the local conditions determine what varieties shall arise, natural selection *creates nothing*. I find this latter point difficult to keep always steadily before the mind.[46]

Bates had not fully mastered the concept of natural selection, even though he had given a beautiful example of it in print. He confused the determination of variations with their selection. The variations Bates was dealing with were principally the local varieties of single species. Throughout the Amazon forest, the life conditions were nearly constant, and yet within the area divergence was observable in geographically localized changes of wing patterns. The alterations hardly seemed to be necessitated by the struggle for life. Since the subtle changes in the local populations did not seem to be part of the life or death struggle, natural selection did not suffice as an explanation. Bates sought an alternative explanation in the influence of the environment.

He felt that there must be subtle differences in the environmental conditions that caused the geographical variants. In turning away from natural selection as an all-sufficient explanation for questions on speciation, he went too far. Hooker was quick to point out the error.

> I hardly understand you when you say that natural selection <u>creates</u> nothing, because you do not give the sense in which you use the word 'create.' I quite agree that natural selection does not explain everything; but it may often <u>suggest</u> an explanation. After all to me the <u>mystery</u> is variation.[47]

Hooker was acutely aware that natural selection was dependent upon materials given to it, that is, variations. Even Darwin had not been able to uncover the mystery of the source and character of variation.

Hooker expanded upon his view that variation is independent of local circumstances in a later letter: "Local circumstances are the secondary causes acting through natural selection on the varieties, killing some, sparing some, directing intermarriages, etc., etc."[48] The difficulties of explicating the action of natural selection without a more precise understanding of inheritance bursts out of these letters. The Darwinians could explain how species change is effected through natural selection based on variations, but they were forced to begin with a mysterious, unexplained "given" in variation.

Hooker and Bates were working out puzzles of comparable

complexity; Hooker was studying plants and Bates insects. Each man had observed in his own subject that species which are stable in one set of conditions are often variable in another set. Hooker cautioned his correspondant not to leap too quickly from that observation to the easiest explanation. The change of environmental circumstances ought not to be construed, Hooker contended, as causing changes in the next generation of offspring. Hooker believed that what had long passed for environmental causation could better be explained by natural selection. He wrote, "Natural selection will act on an imperceptible difference--it searches where no faculty of man can follow."[49]

Hooker was a faithful anti-Lamarckian. He believed that natural selection alone, without the action of any other force or mechanism of change, could produce any degree of divergence from its ancestral stock. In his opinion, even Darwin had underestimated the subtlety and the power of natural selection. Hooker wrote to Bates:

> Darwin, I believe, holds with you as to the influence of external conditions on the variation of the brood. I have, however, failed to be convinced by him of it, and I do not think he recognizes the facts of variation to the extent I do. Indeed I think his book would have been more convincing had he treated variation somehow so as to have impressed the unaccustomed reader and thinker to regard it as the origin of species and natural selection as the fixer of these.[50]

Hooker knew perfectly well that it was not only "unaccustomed" readers" who were subject to misapprehensions about natural selection; even his correspondent Bates had struggled to understand the relation between variation and the fixation of variation by natural selection.

Hooker realized, perhaps better than Bates, the very long periods of time required for variation and natural selection to produce new species. Unless a very long time span is allowed for gradual speciation, evolution seems to demand a directing agent or force. The direct or indirect action of physical conditions were possible ways for nineteenth-century naturalists to explain the appearance of direction in variation. Hooker believed that time was available for natural selection to work on random variations, though he was willing to acknowledge that other causes of organic change might have to be explored in the future should natural selection's action need supplementation. "At present I do not see one or another, and so feel inclined to renounce any _other for the present_."[51] Hooker recognized that this staunch commitment to natural selection was still a relatively new position for him. He continued, "It is not so long since I thought differently, and that variation was the first effect of circumstances on the individual."[52]

The strength of Hooker's arguments against any appeal to direct influence of the environment and against the inherited effect of use and disuse persuaded Bates to contract slightly his reliance on those causes. Bates admitted that these ancillary causes may not be strong enough even to produce a local race without the aid of natural selection. But Bates was not prepared to fully renounce the inheritance of acquired characters. He retained Lamarckian explanations as an option although he was not able to give many examples of their actual occurrence. The case of the mimetic butterflies, for example, exemplified the action of natural selection. Bates admitted in his letter to Hooker that the butterflies illustrated Hooker's stance on natural selection much better than his own.

> A little more corroboration of your views on testing them by my case of the mimetic butterflies, I find they hold good. There is really no proof that the variable species originally sported in a different manner in one locality from what it did in another. Inappreciable variations or very small variations similar in all the localities, and natural selection <u>drawing them out</u>, as it were, would be sufficient to explain the whole case.[53]

Bates had used that same expression of variations being "drawn out" by natural selection in his "Contributions" paper in the *Proceedings of the Linnean Society*, where he had tried to convey the idea that natural selection can work consistently in a linear, directional manner. Natural selection was linear in the sense that it preserves particular adaptive traits in the same manner that the horticulturalist preserves variations in the direction of desirable traits over many generations. The mimetic butterflies were good examples of that kind of long-term directional change. They show the drawing out of the traits of resemblance to a model to their fullest expression.

Bates could not renounce the inheritance of acquired characters totally because Darwin seemed to have made a convincing case for their inheritance. Bates was persuaded that studies on domestic animals had proved the effect of use and disuse on the heredity of succeeding generations. Bates told Hooker, "Mr. Darwin's case of tame ducks having less developed wing bones from disuse seems to come under this head."[54] Darwin's authority seems to have been the major factor in Bates's resolution that acquired characters may be inherited.

The inertia of surrendering a firm belief may have been a factor in addition to Darwin's authority which held Bates back from the belief that natural selection was the only mechanism of species change. The repeated correlations between certain environments and typical characters suggested some form of environmental determinism of animal species. The example Bates mentioned was that seacoast environments are typified by succulent plants and by beetles which tend toward brassy, iridescent colors. It would seem, then, that there must be some way in which those characteristics are induced by the coastal location. If there were no direct or indirect effect of the habitat the succulence and the brassy coloration of the beetles would be found less consistently. Bates acknowledged, "Some part of this general similarity is owing to natural selection adapting one species to another in external dress, but I think there are some cases where the explanation is inapplicable."[35] The effect of the environment seemed to be called for in those cases where no adaptive advantage for characters could be discovered. The systematic appearance of traits, like brassy coloration in coastal beetles, was too constant and not linked to any known utility to be confident that it had resulted from selection. It seemed far more likely to have been produced somehow by environmental agency. Bates confessed, "There is no new view in my line of argument--it is the same as Mr. Darwin's." He continued:

> There is no irreconciliable difference, I think, between your view and his, but Mr. Darwin's seems to harmonise best with the facts of variation as gleaned from the distribution of closely-allied species and varieties which have come under my notice.[56]

The fact that a species in one location remains stable and in another becomes variable seemed analogous to Darwin's findings in his work on domestication, which indicated that the domestic state makes species unstable.[57] Bates also accepted Darwin's opinion that diseases may become hereditary. The possibility of the inheritance of an acquired trait is only revealed in domestic animals since an individual which inherited a disease acquired by its parents would almost surely perish in the wild. Nonetheless, the inheritance of acquired characters cannot be dismissed as a possibility *a priori*, or merely on the grounds that it is not confirmed in wild species. In Bates's work, as in Darwin's, natural selection works toward adaptation on the basis of heritable characteristics whether they are "Darwinian" individual variations or "Lamarckian" acquired characters. Heredity was one subject and adaptation another in Bates's thinking, and he did not always seem clear on the relationship between the two.

Although Bates expressed his confidence that beneficial variations were preserved by natural selection, it was often difficult to know whether environmental conditions actually produced changes in organisms or acted indirectly as the agent of selection. Bates tried to explain how arctic animals had evolved their thick white coats of fur or hair. It was a problem that combined direct action of the environment with natural selection. It was known that sheeps' wool was affected by temperature which alters the structure of the hair. On the basis of that fact, therefore, it seemed logical to Bates that the thickness of the wool was environmentally induced and rendered heritable; but the white coloration was selected by natural selection in precise analogy to the sheep breeder's selection of white sheep in a domestic herd.[58] The examples from domestication were drawn from Darwin's research, not based on Bates's own first-hand experience with animal species.

The series of letters was concluded by one from Darwin to Hooker which was written as a cover letter when he sent on to Hooker the letters between Bates and Hooker. Darwin was absolutely correct when he wrote:

> You have, I think, said all that can be said against direct effects of conditions, and capitally put. But I stick to my own and Bates's side...It is really curiously satisfactory to me to see so able a man as Bates (and yourself) believing more fully in Natural Selection than I think I even do myself.[59]

Of the three, Hooker relied on natural selection most and Darwin least. All three of the correspondents had gained information from the exchange of letters, although only one person stated any shift in his position with respect to natural selection. Bates had moved a little closer to Hooker's position, that natural selection is the exclusive mechanism governing the origin of species. Darwin gained a number of new examples of variation, and Hooker had exercised himself in the art of philosophical argumentation. Hooker's role, as well as Bates's, in the articulation of the theory of evolution by natural selection is deserving of more careful study than it has been given in most histories of the period.

All three of the writers were generalizing far beyond observational data. Darwin's domestic breeding researches had made him vitally aware of the vast amount of variation in single species. He was also prone to apply the information obtained from domestic animals to wild species by direct analogy. The analogy of data from domestic to wild species seems to have made him less ready to believe with Hooker that only natural selection determines species characteristics. Hooker was, also, going beyond any observations, but he was out on a different hypothetical limb from either Darwin or Bates. Hooker was committed to natural selection as the mechanism of adaptive speciation even if he must use the working assumption that the changes selected are in response to environmental conditions too complex for human understanding. Hooker was not disposed to add what seemed like gratuitous additional factors when natural selection alone was adequate. Perhaps the most surprising element in the correspondence was that Bates was encouraged not to rely only upon natural selection by Darwin. Even though Bates's personal observation tended to suggest Hooker's position, he held to Darwin's position that "natural selection does not explain everything."

The correspondence on theoretical questions was suspended during the summer and fall of 1862, probably under the pressure of other obligations. Bates was spending his full time preparing the two volumes of the Naturalist on the River Amazons.[60]

Natural Selection in a Popular Travel Book

Bates's paper on the mimetic butterflies had thoroughly impressed Darwin. He liked Bates's way of arguing and the mixture of fact and theoretical analysis that he could bring to bear on the subject of evolutionary change. In April of 1861, Darwin had occasion to write to Bates asking for information on the domestic breeding practices of the tribes he had encountered in the Amazon. In the course of the letter, he encouraged Bates to publish more fully on the mimetic butterflies and to publish a book of travels.

> I hope you will not think me presumptuous in saying how much I have been struck with your varied knowledge, & with the decisive manner in which you bring it to bear on each point,--a rare and most high quality, as far as my experience goes.--I earnestly hope you will find time to publish largely: before the Linn. Soc. you might bring briefly out your views on species.--Have you ever thought of publishing your travels and working in them the less abstruse parts of your Nat. History? I believe it would sell, & be a very valuable contribution to Nat. History.[61]

Darwin's encouragement was just the kind of incentive Bates needed to keep going on the book he planned to write. The writing went slowly and by June of 1862 it was beginning to wear on his patience. He wrote, "My book 'progresses' it is true, but slowly...I am at it every day as many hours as I can stand: but what takes me one day to write, takes 5 to alter."[62] The product produced was fully worth the labor that went into it. Darwin provided consistent moral support with short letters expressing interest in the project, and he also read the first few chapters of Bates's manuscript in draft form.[63]

Darwin's assessment was entirely positive; he wrote: "My criticism may be condensed into a single sentence, namely, that it is the best work of Natural History Travels ever published in England."[64] That was particularly high praise since he obviously did not exempt the _Journal of Researches_ describing the voyage of the _Beagle_. Just as Darwin had reviewed Bates's "Contributions" paper in the _Natural History Review_ to guarantee that it not be overlooked, he also reviewed the _Naturalist_ in the same journal. Darwin proclaimed that the book was "One of the most entertaining and instructive books of travel it has ever been our good fortune to peruse."[65]

The book met with enough success to call for another edition. A one-volume abridgement of the _Naturalist_ was published in 1864.[66] The first edition contained many more natural history facts than the second did. The first edition contains Bates's discussion of insect coloration in the first chapter.

Bates asserted at the very beginning of his book that

the traditional notion that tropical insects are brighter in color than temperate insects is false.[67] Bates had tested that belief by choosing two insect groups having representative genera in England as well as in South America. The groups chosen were the tiger-beetles and the butterflies. (Cicindelidae and Lepidoptera). He had made the rather surprising discovery that the Amazonian tiger-beetles were smaller and of less notable coloration than their English relatives. Of the eight butterfly genera common to both places, three were more beautiful in their European representatives by human tastes, three more beautiful in South American forms, and two were nearly alike in respect to size and brilliance. If climate was the most important factor for understanding insect colors and forms, one would expect the climate to influence both sexes of a species equally. Observations opposed that expectation, however. Bates had found that a very pronounced sexual dimorphism was one of the most notable features in the South American butterfly fauna. Bates suggested to his readers that the key to understanding the coloration of insects was not so much to be sought in their relationship to climatic factors as in the relationship between the two sexes of the species. Although correspondence during the writing of the _Naturalist_ indicates that Bates was holding open the possibility that climate could directly influence organisms and shape their heredity, he did not emphasize that in the Naturalist. In the book of travels, he discussed the need for adaptations for survival, and their gradual development by natural selection.

Bates knew the audience for the Naturalist on the River Amazons was different from the readers of the Proceedings of the Linnean Society. In the Naturalist, his descriptive capacities found full expression in unfolding the details of exciting and dangerous adventures among anacondas, crocodiles, and finch-killing spiders. Bates was master of the literary milieu as well as of the scientific. He wrote vivid accounts of the lives of exotic colonial insects and trenchant passages on the meaning of insect coloration, and particularly mimicry in butterflies. Passages on the origin of species and a few geographical remarks were paced so as not to weary the reader who opened the volume seeking an adventure story with a modicum of scientific discourse. Bates succeeded as few had at satisfying both the popular and the scientific audience.[68]

Only a few really new facts were presented in the Naturalist which had not been published before. Many of the stories had appeared in his letters to The Zoologist or in one or another installment of the "Contributions to an Insect Fauna of the Amazon Valley," but only a small and special audience of naturalists would have seen them. Although the facts were not new to everyone, the explanation of how natural selection works to make new species would have been new to many readers of popular literature. Bates described what it was like for the travelling naturalist to be in daily contact with geographically varying species which he thought could best be understood on Darwinian principles. Altering his language slightly, Bates told his general reader what he had told the Fellows of the Linnean Society in 1861. "It is scarcely necessary to add that the conclusions thus arrived at will apply to all organic beings."[69] In a characteristically Batesian understatement, he told the readers that evolution occurs and that if his conclusions are correct with regard to South American insects, they apply also to all other creatures. Emma Darwin, who certainly qualifies as a "general reader," said that Bates wrote with "so strong an air of truthfulness."[70]

In many ways, the Naturalist and its warm reeeption were the high points of Bates's career. It provided him the forum for a personal statement of his goals as a collector, as an entomologist, as an evolutionist. The Naturalist was the conclusion for Bates of the commitment which Alfred Wallace had framed and Bates had seconded in 1847 before their departure for the Amazonian forest. Their purpose in undertaking the journey was "to take one family to study thoroughly, principally with a view to the theory of the origin of species."[71] Bates's study of butterflies had led him to new understandings of adaptation, variation, and speciation.[72]

He had seen what he believed were new species forming from varieties through the process of natural selection and he was able to state it clearly and beautifully.

Near the end of the book he summarized what the study of butterflies had meant for him.

> But no description can convey an adequate notion of the beauty and diversity in form and colour of this class of insects in the neighborhood of Ega. I paid especial attention to them, having found that this tribe was better adapted than almost any other group of animals or plants to furnish facts in illustration of the modifications which all species undergo in nature under changed local conditions...It may be said, therefore, that on these expanded membranes [butterfly wings] Nature writes, as on a tablet, the story of the modifications of species so truly do all changes of the organisation register themselves thereon. Moreover, the same colour patterns of the wings generally show, with great regularity, the degrees of blood relationship of the species. As the Laws of Nature must be the same for all beings, the conclusions furnished by this group of insects must be applicable to the whole organic world: therefore, the study of butterflies--creatures selected as the types of airiness and frivolity--instead of being despised, will some day be valued as one of the most important branches of Biological science.[73]

The concept of the unity of nature and natural laws was not new with Bates. The earlier naturalists who spoke out of a framework of belief in natural theology also ascribed oftentimes to the idea that the "Laws of Nature must be the same for all beings." That position was often asserted when biological analogues to the Newtonian concept of gravitation were given. Bates had departed from the older concepts of the unity of nature stemming from a common origin in the creative intelligence. The unifying force he had seen in his studies of butterflies could bear extrapolation to all other groups. Nature had sketched the tale of evolutionary modification on the wings of butterflies where the coded message could be deciphered.

Despite the general acclaim and good sales of the Naturalist on the River Amazons, Bates found it difficult to get good employment where he could do original research. He was not likely to be hired at the British Museum, where Richard Owen and John Edward Gray were keepers of collections and opponents of Darwinian ideas. Bates balked at undertaking taxonomic work for private collectors, since it drained him of the energy he needed for original research. Now that he had a successful book, he was the object of the envy and hostility of established entomologists. John Edward Gray, who was extremely powerful as the Keeper of the Zoological Collections at the British Museum, castigated Bates for overestimating his contributions of new species. Bates had stated that 8,000 of the 17,700 species collected were new to science. Bates was shocked at the vehemence of Gray's negative reaction to him.[74] Bates told Hooker in a letter that he had fallen into a "nest of hornets." Hooker did his best to comfort Bates; he told him that "It is all along the law of the struggle for life..."

> Let them suppose they will carry their mole-hill, and do you take the mountain...Above all things remember that entomologists are a poor set, and it behooves you to remember this in dealing with them.[75]

Gray was a stereotype of the professional who has both social and occupational prestige, and he was unable to be generous to men who lacked professional standing. Wallace, like Bates, found it difficult to break into London scientific society. Both travelers were respected for their scientific work, but they were self-taught and therefore permanent amateurs. Neither man was ever able to find long-term employment in science which provided financial security and opportunity to do original research. Bates, whose claim to 8,000 new species was eventually vindicated, wrote sadly:

> The hardships of foreign travel have the good effect of cooling the ardour of one's expectations and ambition. I do not know whether it is the same with other travellers as myself, but these tough experiences seem to bring one under the truest of the beatitudes, namely, "Blessed are those who expect little or nothing, for they shall not be disappointed."[76]

Protective Resemblance in Later Monographs

Bates disliked intensely the bitter antagonism he had found in the closed society of established entomologists. His later writings are more taxonomic and less theoretical than his earlier publications. It may have been that this was

Bates's way of withdrawing from the antipathies in entomology at that time, but without denying entomological science his positive contributions as a systematic zoologist. Bates had promised a paper to the Natural History Review which was to be a survey of the whole subject of geographical relations of species and varieties. He laid that theoretical effort aside in preference for classificatory work on the collection owned by W. W. Saunders.[77]

Unquestionably, the fact that many of the specimens he was cataloguing were those he had collected himself contributed to his enjoyment of the work. Bates admitted that he had a tendency to overindulge his preference for "naming, arranging, labeling, etc.--when other more important matters are waiting."[78] It seems likely that Bates was motivated by a combination of positive and negative factors. He stated positively that "The monograph is so much pleasanter work just now that I have laid this paper aside for the present."[79] The taxonomic work enabled him to avoid becoming embroiled in the controversies a more theoretical paper would have involved.

Bates liked to work up to theoretical problems inductively, through the presentation of a large volume of data about organisms rather than to confront matters of theory directly. His study of W. W. Saunders collection of Phasmidae not only introduced fifty-two new species, but it surveyed the relationships of the species to their ecological and geographical situations.[80] He retained the eye of the ecologist-biogeographer even when he had turned to the efforts of the taxonomist.

The Phasmidae are the family of Orthoptera that are called "walking-stick insects." They generally resemble some part of plants such as twigs, leaves or bark. Like the Heliconiidae, their resemblance is a protective adaptation. The analysis of the Phasmidae required Bates once again to come to grips with the philosophy of taxonomy in conjunction with his belief in their gradual evolutionary origin. These insects were also "mimics," but this time they resembled plants, not other members of their own order.[81]

The geographical picture of Phasmid species confirmed Bates's belief that their classification had been based on artificial distinctions. Taxonomists previously had used the structure of the wings as the key character for distinguishing species in the Phasmidae. Bates believed he had found the walking-sticks with wings were frequently very closely related to those without wings, and vice-versa. Nature had not placed Phasmids with wings in one genus and those without wings in another. That was a man-made dichotomy not found in nature.

An alternative to the taxonomic scheme based on wing structure had been suggested by Adolf Gerstaecker, which used the armatures of the legs as its key character.[82] Bates believed that both the wing structure system and the leg system were partakers in the same erroneous belief that the group had been endowed by nature or by God with certain generic traits. Bates had long given up the hope that he, or any other monographer, could retrace the work of nature, since nature had failed to endow the Phasmids with constant traits.

> I believe that in the Phasmidae, as in some other groups, Nature, in adapting her species to their conditions of life, has, in the process of variation and adaptation, involved all those parts of structure which usually yield, by their partial constancy, characters for the definition of generic groups.[83]

When nature leaves no character constant, the taxonomist must do the best he can, admitting that his single character classifications are artificial. What was wrong in taxonomy was the creation of an artificial system with the false pretense that it was a natural system.

Bates explained why some families were easy to cast into genera and why some like the Phasmidae were so extraordinarily difficult. He made the distinction between heterogeneity and homogeneity of type.[84] Organisms which were readily classified were heterogeneous in type. He meant by that term that there had been an early divergence into several modes of life in a group sharing common ancestry. The distinct modes of life permitted the evolution of specialized structures through natural selection. Animals of heterogeneous type were those having both specialized structures and some organs which were relatively constant, which connect it to its relatives. The stable structures can be used by taxonomists as good generic criteria. The Phasmidae were not, however, heterogeneous in type.

All the walking-sticks had approximately the same habitats and niches. Nature had left no structure unaltered to aid the taxonomist in his attempt to divide them into several genera. Since there had been so little ecological divergence, the taxonomist was faced with a dilemma. Should they remain in an unwieldy and dubious single group, or should an artificial system be employed? Bates believed that the study of Phasmidae was another illustration of his long-held position that believing in evolution made the work of classifying large groups of organisms much harder. He did not have the aid of the assumed belief that he was discovering the natural scheme the Creator had used in the manufacture of permanent species. Bates was of the opinion that if the type of the group is homogeneous, the taxonomist must adopt a pragmatic approach. Since, in the case of the Phasmidae, he had not been able to identify any character or characters which would give a more natural classificatory system than Westwood's system, based on the structures of the wings, Bates adopted the alary system. Westwood had worked out his taxonomic scheme for the Phasmidae incorporating the researches of the earlier workers such as Lichtenstein, Gray, Burmeister and DeHaan.[85] A natural system for an evolutionist was a reconstruction of the historical divergence of species, but when it was not possible to make such a reconstruction, Bates was willing to employ the best approximation of it he could find.

Besides his discussion of the systematics of the Phasmidae, Bates had studied thoroughly the various adaptations for defense found in the group. These adaptations gave him insight into the large variety of evolutionary alternatives available to organisms who had to solve a given problem. In commenting upon adaptations for camouflage, in which one walking-stick resembled a narrow blade of grass and another a twig, and yet another a leaf, he wrote: "All these properties, it is needless to remark, are the varied means by which these apparently helpless creatures succeed in maintaining their place by the side of the nimbler, and more compactly organized families of insects."[86]

Most Phasmids had had adaptations for hiding rather than for active escape through speed. He had studied the strange process in the Heliconiidae of hiding by means of conspicuous coloration and in the Phasmidae hiding was accomplished mostly by cryptic coloration, but there were a few forms that were not plain shades of brown or green.

> Some species of *Phasma* and *Necrosoia* (winged in both sexes) have no disguise, but, on the contrary, brilliant and varied colours: what means of protection they enjoy is unknown, but the fact shows the justice of the view that these are typical Phasmidae; they have not been perverted by the agencies of special adaptation into a resemblance to a stem of grass or a leaf.[87]

Protective coloration once again provided opportunity to test the evolutionary substructure being used as a framework for the taxonomy of the Phasmidae. Bates did not know what function the conspicuous colors served, but he believed that they were adaptive or they would not have survived the operation of natural selection. Bates conceived that these brilliantly colored *Phasma* species were close to ancestral type; that they had not diverged in the direction of one plant part or another. Even the occasional exceptions to the general rule that walking-stick insects have cryptic coloration illustrated the action of natural selection. The descriptive papers Bates wrote illustrated the workings of natural selection, but only to those inside entomology. Bates was much more the archetypal Baconian in that he could only reason about theories in the context of particular data.

Darwin had always claimed that he worked on "Baconian Principles," but he was always far more willing than Bates to abstract his theory from its corroborating data.[88] Indeed, Darwin usually had his theories first! Bates, however, would never divorce data from theory, and it had both beneficial and deleterious effects on his work and his career. Bates's observations were a great boon to Darwin and Wallace, who put particular facts into more general context within the evolutionary theory. Bates was a respected taxonomic authority not only on the butterflies and the walking-sticks but also on beetles. His work was greatly admired and of permanent value. At least one committed Darwinian was able to make his way as a systematic entomologist without compromising his Darwinian principles. The orientation toward observational data in his geographical work and toward taxonomy in his later work had negative influence on the assessment made of Bates in the few historical works that discuss his accomplishments.

Only one book-length biography of Bates has ever been published, and it shows little understanding of what Bates was doing in the years following his Amazonian travels. George Woodcock's biography _Henry Walter Bates, Naturalist of the Amazons_ ends with the following expression of bewilderment.

> Yet one is still at a loss to determine why the young man who went to the Amazons with such zeal and who returned so hopefully to lay his ideas before the world, should have come to such a modest end.... For four years, after his return, the stimulus persisted, but when Bates finally decided not to travel again, and gave hostage to fortune by marrying, the impulses born on the Amazons diminished and lost their power to inspire him.[89]

It must be noted that Woodcock does not mention Bates's careful taxonomic studies that incorporate evolutionary concepts. Bates had a penchant for hiding his light under a bushel by choosing such titles as "Contributions to an Insect Fauna of the Amazon Valley" and "Descriptions of Fifty-two New Species of Phasmidae from the Collection of Mr. W. Wilson Saunders, with Remarks on the Family." It must be understood that Bates accorded the "facts" the first place and his title was a way of stating a priority. In the Heliconiidae, the phenomenon of mimicry was found in the process of studying geographical distribution and it remained in that context when he reported it. He learned much about adaptation and natural selection in the Phasmidae while he was in the process of studying them for W. W. Saunders. The theoretical observations remained in their original context when he published his paper on the Phasmidae.

It is a common bias in historical studies to treat only innovations in theories. Woodcock did not mention Bates's paper on the Phasmidae or his later writings in any way that indicates that he actually studied them. Taxonomic papers have been treated as unimportant. The bias against systematic papers has tended to obscure the contributions which Bates made after the discovery of mimicry, since it was only then that Bates began to find ways to turn the taxonomic enterprise into evolutionary phylogeny.

Bates did not drop out of the Darwinian debate. He pursued his own private vision of what was most important to it. The later taxonomic monographs were expressions of the declaration Bates made to Darwin in one of the earliest letters they exchanged, in 1861. Bates then stated:

> I am quite convinced that insects offer better or clearer illustrations of the problems you occupy yourself with than any other class of animals or plants. It is so easy with them to obtain great series of examples and have them before you in a small compass, which is one advantage they have.[90]

The problems Darwin was interested in, in 1861, were variation and sexual dimorphism. Bates's study of mimicry and of defensive adaptation in general in relationship to natural selection were ideal for inclusion in the later versions of the *Origin of Species*.

Batesian Mimicry in the *Origin of Species*

Darwin introduced the subject of Batesian mimicry in butterflies in the fourth edition of the *Origin of Species*. and retained it in succeeding editions.[91] It is interesting that Darwin discussed mimicry in chapter 13, in the section on classification, where he also discussed adaptive and analogical resemblances in contrast to the resemblances which result from genetic affinity. That was precisely the usage Bates had made of the discovery in his own "Contributions" paper. Darwin might have used the information in several other ways. For example, he might have used mimicry to illuminate the laws of variation, or the procedure of natural selection. Since Darwin appreciated the force of Bates's paper relating taxonomic questions and evolutionary ones, Darwin preserved the same context when he discussed mimicry in the _Origin of Species_.

The subject of analogical resemblances had been broached by Lamarck and elaborated by William MacLeay many years before.[92] Many of these analogues or repetitions in nature were already well known: for example, the resemblances between the fins of fish and of whales, and between the prosimian shrew-mouse and the rodent mouse. The phenomenon is now called convergent evolution, and although Darwin did not use that term in his section on analogical resemblance, he employed the concept that diverse means may be used to gain single ends. He was very careful to separate cases in which resemblance could be attributed to common inheritance and cases in which genetically very distinct animals responded to the challenges of survival by accumulating adaptive variations which led to external similarities. Darwin stated in the sixth edition of the _Origin_ that animals only distantly related are likely to produce similar variations, "And this would obviously aid in the acquirement through natural selection of parts or organs, strikingly like each other, independently of their direct inheritance from a common progenitor."[93] Most of the parallelisms seen in nature were attributable to the preservation of adaptations to similar modes of life, but Bates had found a new form of analogical resemblance, "In which close external resemblance does not depend on adaptation to similar habits of life, but has been gained for the sake of protection."[94] Darwin reminded his readers that these mimetic analogies exemplified the survival of the most convincing mimics and the elimination of the imperfect. Given a long period of time allowing variations in color to occur, an unprotected butterfly species could gradually come to bear an external visage deceptively similar to a protected model living in its area. Darwin stated, "Here we have an excellent example of natural selection."[95] Mimicry would be most perfect where the senses of predators were most acute.

Darwin wanted to demonstrate that natural selection could make very large changes in animals since it involved preservation as well as destruction. The mimics were different from their nearer relatives in color, pattern and also in the shapes of their wings. Also, mimicking butterflies frequently behaved differently from their relatives. Bates's discovery of mimicry and his evaluation of its meaning gave Darwin an excellent example to use in his explanation of the effect of natural selection in causing divergence of species from their close allies.

Mimicry in butterflies was a good point. Also, for Darwin to discuss the "social context" of evolution, that is, the interdependence of species. The mimics gained survival as a species by their deception, but they sacrificed their independence. The geographical range of the mimic was limited by that of the model. The population of the mimics must be only a small percentage of the population of the model. Darwin wrote that one of the "general lessons learned from the study of mimicry is that speaking metaphorically they [mimetic insects] are reduced, like most weak creatures, to trickery and dissimulation."[96]

The disguises of nature which are analogical in nature demonstrated the lengths to which natural selection could go in preserving the favored races in the struggle for life. During the period of Darwin's revisions of the <u>Origin of Species</u> for editions four, five, and six, new data were becoming available to him which made Batesian mimicry a general and world-wide phenomenon. It is, therefore, a bit curious that Darwin's strongest expressions of delight occur in the earlier rather than in the later editions when he had corroborating testimony from Alfred Russel Wallace and Roland Trimen. Darwin introduced the new discovery in the fourth edition of 1866 as follows:

> The most remarkable case of analogical resemblance ever recorded, though not dependent on adaptation to similar conditions of life, is

> that given by Mr. Bates with respect to certain butterflies in the Amazonian region closely mimicking other kinds.[97]

The statement is a much livelier introduction to a phenomenon so perfectly suited to explain the _modus operandi_ of evolutionary descent with modification than the statement that replaced it in the sixth edition of 1872. Darwin stated simply:

> I allude to the wonderful manner in which certain butterflies imitate, as first described by Mr. Bates, other and quite distinct species.[98]

No motive for the change was given either in print or in published letters. It may have had something to do with the fact that Darwin had included material on mimicry in the _Descent of Man_, where it is used to argue for sexual selection rather than natural selection.[99] Darwin, of course, may have come to believe that analogical resemblances had been found subsequently to Bates's discoveries that were more remarkable. If that were Darwin's reason for changing his introduction, he was quite correct that Roland Trimen's discoveries were visually much more notable, although they were of Bates's genre of mimics.[100]

Bates had been an important resource person for Darwin during the revisions of the _Origin_ and the writing of the _Descent of Man_. His competence on technical entomology stood Darwin in good stead on the subject of sexual coloration in Lepidoptera. Justifiably, Darwin was grateful to Bates for his offer to read the draft of chapters for the new book:

> Your most kind offer of reading my MS on Lepidoptera has made all the difference to me: if I had not known this I shd have been tormented the whole time with doubts. I have tried to be cautious and to avoid blunders, but when one writes on a subject, with which one is not familiar, mistakes inevitably occur. P. S. Two-thirds of the MS relates to facts communicated by you to me, or published by you.[101]

It is very likely that Darwin exaggerated the proportion of facts Bates had contributed directly. Nevertheless, Darwin fully realized what an asset Bates's work had been to his own; he had been repaid many-fold for the work he had done in reading Bates's <u>Naturalist on the River Amazons</u> before its publication. Darwin felt he could rely on Bates to keep him from erring in print. Published letters from Bates to Darwin show that Bates made suggestions for alterations in the orthography of insect nomenclatures in Darwin's manuscript and he offered some additions to Darwin's stock of examples of sexual dimorphism in Coleoptera.[102] Perhaps the most helpful suggestion Bates made was that Darwin bolster his presentation of facts on beetles, which was inadequate to carry Darwin's argument.

Bates may have been Darwin's choice as a collaborator because they were in agreement on all important theoretical points. Unlike Wallace, Bates had no strong reservations or objections to the theory of sexual selection. In general, Bates shared Darwin's opinion that it was generally in the male sex that innovations in color occurred, the male sex generally being more greatly differentiated from the ancestral form than the female of the same species. The same data was used by Wallace to quite opposite conclusions. The same colors Darwin and Bates called sexual colors, accumulated by sexual selection, Wallace called protective coloration accumulated by natural selection. Although there was no animosity between Darwin and Wallace, Bates hoped that a synthesis might be found.

As a friend and correspondent of both Darwin and Wallace, Bates would have liked to catalyze the reconciliation of the two views on sexual dimorphism. In a letter accompanying the packet of Darwin's manuscript, Bates wrote:

> I almost always agree with you & have less scruple on this account to suggest a modification of your views of females not being more dull-colored by selection which will bring your opinion & Wallace's nearly into harmony. It is this--the necessity of females being dull-coloured for protection is true, but they have not been _made_ dull from former brighter hues, but have simply been _kept_ dull by natural selection steadily eliminating all tendency to brightness. This will not disagree with your clinching & true argument against Wallace that females of a genus are truer in colour to the generic type than males.[103]

Bates's assistance in helping Darwin think through his arguments was always welcomed by Darwin. In the work for the _Descent of Man_ alone, Bates provided Darwin with information on the sex ratios of insect groups, on sex differences in insects, on sexual dimorphism in mimetic butterflies, and on which sex most resembles the generic type.

Bates and the Darwinian Data

Henry Walter Bates is remembered by biologists as the discoverer of the form of mimicry that bears his name, Batesian mimicry. He is also remembered in historical studies of the biology of the nineteenth century as the author of the *Naturalist on the River Amazons*, one of the best travels and natural history books ever written. These are the only accomplishments usually ascribed to Bates. It must be noted that Bates's contributions were seen more broadly during his own lifetime.

The prestige of the subjects which Bates valued highly--taxonomy and biogeography--declined in the period between 1890 and 1970. The success of the Darwinian revolution also served to diminish the memory of many of the early contributors to the revolution. The change in methodology which Darwin's work demonstrates has become the focus for some of the best studies of the important revisions in biology in the middle of the last century. Michael Ghiselin's important book *The Triumph of the Darwinian Method* presents an excellent study of Darwin's views.[104] Something is lost, however, in the abstracting of methodology from its content. It serves to unintentionally denigrate the value of the content. For example, he discusses Darwin's views on mimicry without making clear that Darwin had a vast amount of help in devising a proper statement of it in both the _Origin_ and the _Descent of Man_. Darwin has become the central symbol of the Darwinian revolution as it is proper that he should be. However, many friends of Darwin participated in the provision of data and it is important not to obscure their contributions.

Bates's biographers have nearly unanimously bemoaned the fact that he changed his scientific focus after 1864. George Woodcock wrote:

> From soaring with his friends Wallace and Darwin in the high air of theory, he now concentrated on the very 'cabinet work' which he once felt the pundits of the British Museum over-emphasized, 'naming, arranging, labelling, etc.', and extending these activities into elaborate tasks of classification, which were undoubtedly necessary, but which could have been done by a man of lesser capacity.[105]

It would seem unlikely that Woodcock would have made that statement had he read Bates's study of the taxonomy of Phasmidae, for example. A man of less capacity would have classified the group according to established principles and missed completely the implications in the group for the evolutionary theory. There is much more to taxonomy than arranging and labelling. Bates knew very well that it was important for an evolutionist with insight into the gradual origin of species and their flexibility in response to different ecological conditions to do the classifying. The combination of insights from biogeography and from taxonomy made it possible for Bates to make lasting contributions.

Bates was one of the first to apply the new theory to practice. His work is marked by his sensitivity to the gradual emergence of species from varieties. Undoubtedly he did not have opportunities that permitted the full flowering of his potential as a theoretician and philosopher. There are elements in his personal psychology as well as in his later career as Assistant Secretary of the Royal Geographical Society that determined his choice to do the concrete kind of work taxonomists do instead of the abstract work of the theoretician.[106] Theories are not the only important contributions, however. And even in theories, Bates had a greater behind-the-scenes role as a correspondent of Darwin, Hooker, and Wallace than has generally been acknowledged. He was a provider of Darwinian data who had a singularly practical and constant vision of what needed to be done if evolutionary theory was to gain an important place in the descriptive sectors of biology. His later work is better seen as an extension of his earlier studies instead of a denial or repudiation of it.

Bates's letter to Darwin before the publication of the "Contributions" paper, giving the details and theory of mimicry in butterflies states his permanent conviction:

The group from which I have supplied this example [of species arising from varieties] is very interesting,--being restricted to S. America, which swarms with its species; & apparently a modern creation. I am now studying it with a view to writing upon it. I think I have got a glimpse into the laboratory where nature manufactures her new species.[107]

Notes--Chapter II

BATES AND THE LABORATORY WHERE NATURE MAKES HER NEW SPECIES

1. Henry Walter Bates (b. Leicester, England 1825; d. London, England 1892) was the son of a Leicester manufacturer of hosiery. He became an amateur insect collector while still a very young man. He met Alfred Russel Wallace in 1844 or 1845. Wallace (1823-1913) was a teacher at that time in the Collegiate School in Leicester. The friendship changed both of their lives. Bates introduced Wallace to entomology and Wallace suggested to Bates that they undertake a trip to some tropical country to collect specimens. They left England for the Amazonian area of Brazil in May, 1848. Bates remained in Brazil eleven years, returning to England in 1859. Wallace returned to England in 1852; during the voyage his ship burned and most of his specimens were destroyed. While Bates was still in Brazil, Wallace set out again for the Malay Archipelago in 1854, and remained until 1862. Neither Bates nor Wallace was able to secure eimployment in science. Wallace supported himself and his family by writing primarily, and Bates worked as assistant secretary of the Royal Geographical Society of London after 1864.

2. The importance of Charles Lyell's (1797-1875) work on the history of the earth as published in his _Principles of Geology_ (3 vols. 1830-1833) can scarcely be overestimated in its influence on both Darwin and Wallace. Lyell's emphasis on geographical distribution and on the continuity of time and natural forces was adopted as central focus by both discoverers of natural selection as their notebooks show. Some of Wallace's notes have been published by Henry Lewis McKinney, _Wallace and Natural Selection_ (New Haven: Yale University Press, 1972). Darwin's notebooks were published by Gavin DeBeer in _Bulletin of the British Museum (Natural History) Historical Series_ (London, 1959-60, and 1967).

3. Bates, "Excursion to St. Paulo, Upper Amazons," _The Zoologist_, _16_:6160-6164, 1858 (hereinafter called "Excursion.") Although this is Bates's first published account of the relationship between _Leptalis_ mimics and _Ithomia_ models it is the second paper which features the _Ithomia_. They are one of the genera of the Heliconiidae. The first is "Notes on South American Butterflies," _Transactions of the Entomological Society of London_ _4_:1-11, 1859-61. Read

September 7, 1857. Bates was working without the benefit of plates to demonstrate the wide departure necessary for a species of *Leptalis*, a member of the family Pieridae, to resemble a member of the family Heliconiidae. The resemblance was produced in the *Leptalis* butterflies by substantial changes in body parts in addition to coloration patterns. Bates suggested that taxonomists not rely too heavily upon the nervures of wings as taxonomic criteria which the taxonomist William Hewitson considered reliable guides to discriminating closely related species.

4. Bates, "Excursion," p. 6165.

5. Bates, "Excursion," p. 6165.

6. Bates,"Excursion," p. 6166.

7. Bates, "Excursion," p. 6166. The nearly exact duplication of the colors of a moth which was found in a butterfly was surprising enough, but the moth mimics seemed to respond as readily to changes in models as butterfly mimics did. For example, the Bombycide moth, which resembled an *Ithomiae* at Fonte Boa made the appropriate changes to parallel the modifications in the *Ithomiae* at Ega and at St. Paulo.

8. The "Diurnes" are the diurnal lepidoptera and a colloquial synonym for butterflies which was used to distinguish them from the parts of the order Lepidoptera which includes the moths.

9. Henry Walter Bates, "Contributions to an Insect Fauna of the Amazon Valley. Part I. Diurnal Lepidoptera," *Transactions of the Entomological Society of London* 5:223-228, 335-361, 1858-1861. Pages 223-228 read March 5, 1860; pp. 335-361 read November 24, 1860. Genus *Papilio* is a member of the family Papilionidae, commonly known as the swallow-tailed butterflies.

10. Charles Darwin, *On the Origin of Species* (London: John Murray, 1859). The edition used here is the facsimile of the first edition with an introduction by Ernst Mayr (Cambridge, MA: Harvard University Press, 1966). Chapter XI, "Geographical Distribution," p. 378. Bates found that in the genus *Papilio*, eighteen species, of the forty he studied, were endemic to the tropics, ranging no more than twelve degrees north or south of the equator. Darwin had concluded that at the coldest period, the climate at sea level in the tropics was about what it is now at six

or seven thousand feet. The tropical lowlands would have been covered with a mixture of tropical and temperate vegetation, "like that now growing with strange luxuriance at the base of the Himalaya, as graphically described by Hooker" (*Origin*, p. 378). Darwin wrote to Hooker: "I cannot give up equatorial cooling. It explains so much and harmonises with so much" (*Life and Letters*, March 27, 1861, pp. 361-362). Bates's *Papilio* were telling their own history. Darwin stated that Bates had him in a "cleft stick," and it seems likely that the force of Bates's arguments served to increase Darwin's respect for his younger colleague.

11. Bates, "Contributions...Diurnal Lepidoptera," 1860, p. 226.

12. Bates, "Contributions...Diurnal Lepidoptera," 1860, p. 342. Bates had taken note of the fact that *P. bolivar* was more geographically isolated from other members of its group than was often found. He believed that the isolation of the species was particularly significant since the species was apparently very susceptible to change. Bates was trying very hard to define the parameters involved in changes of species with some rigor. No one factor, however, seemed to be present all of the time. He often found gradual variation from the center of a species's range moving toward its periphery, but not always. There were cases where it seemed that species changes had been sudden. In these cases Bates assumed that one species had probably arisen from the other with subsequent loss of intermediate forms. Bates considered the influence of the environment as a direct agent of change, but he found that it also was hard to define with precision.

13. Bates, "Contributions...Diurnal Lepidoptera," 1860, p. 354.

14. Darwin, *On the Origin of Species*, p. 489. Darwin was definitely thinking of the temperate English countryside when he wrote: "It is interesting to contemplate an entangled bank, clothed with many plants of many kinds, with birds singing on the bushes, with various insects flitting about, and with worms crawling through the damp earth, and to reflect that these elaborately constructed forms, so different from each other, and dependent on each other in so complex a manner, have all been produced by laws acting around us."

15. Bates, "Contributions...Diurnal Lepidoptera," 1860, p. 354.

16. Bates sometimes used the spelling of the family Heliconidae. I have adopted the modern spelling of Heliconiidae except in quotations. Among the mimics were *Papilio pausanius* which mimics *Heliconia clytia* and *Papilio zagreus* which mimics the Heliconiid genus *Lycorea*.

17. Alfred Russel Wallace, "On the Law which has Regulated the Introduction of New Species," *Annals of Natural History* 16:184-196, 1855; and "Note on the Theory of Permanent and Geographical Varieties," *The Zoologist* 16:5887-5888, 1858. Wallace's letters and notebooks indicate that he could not find the absolute species in nature as different from the permanent variety. See McKinney, *Wallace and Natural Selection*.

18. Letter, Wallace to Bates, January 4, 1858, *Letters and Reminiscences of Alfred Russel Wallace* (London: Cassel & Co., 1916), pp. 53-55. Wallace told Bates that "It was the promulgation of Forbes' theory [of polarity] which led me to write and publish, for I was annoyed to see such an ideal absurdity put forth when such a simple hypothesis will explain *all the facts*." Wallace indicated that he had plans for a book backing up his assertions. "That paper is, of course, only the announcement of the theory, not its development. I have prepared the plan and written portions of an extensive work embracing the subject in all its bearings and endeavoring to prove what in the paper I have only indicated."

19. Letter, Bates to Wallace, November 19, 1856, *Letters and Reminiscences*, p. 53.

20. Letter, Bates to Wallace, November 19, 1856, *Letters and Reminiscences*, p. 53.

21. H. W. Bates, "Contributions to an Insect Fauna of the Amazon Valley. Lepidoptera: Heliconidae," *Transactions of the Linnean Society of London* 23:495-566, 1862. Read November 21, 1861.

22. Bates, "Contributions...Heliconidae," pp. 501-502.

23. Bates, "Contributions...Heliconidae," p. 498.

24. Bates, "Contributions...Heliconidae," p. 502.

25. Bates, "Contributions...Heliconidae," pp. 500-502. Writing on the polymorphic species of the Heliconiidae, *Mechanitis polymnia*, Bates wrote: "We see here the

manufacture, as it were, in process...The external conditions in certain localities are more favourable to one or more of the varieties there existing than to the others; these favoured ones, therefore, prevail over the others" (p. 501).

26. Bates, "Contributions...Heliconiidae," p. 507.

27. William Kirby and William Spence, Introduction to Entomology, Letter XXI. "Means by which Insects Defend Themselves."

28. Cuthbert Collingwood, "On Recurrent Animal Form, and its Significance in Systematic Zoology," The Annals and Magazine of Natural History 6 (series 3) pp. 81-91, August 1860. Bates had communicated to Darwin about his forthcoming paper in the Linnean Society. Darwin responded enthusiastically, "What a capital paper yours will be on mimetic resemblances. You will make quite a new subject of it. I had thought of such cases as a difficulty; and once, when corresponding with Dr. Collingwood, I thought of your explanation but drove it from my mind, for I felt that I had not knowledge to judge one way or another." Letter, Darwin to Bates, September 25, 1861, More Letters of Darwin, Francis Darwin editor,(London, 1903) vol. I, pp. 196-197. (Cuthbert Collingwood, b. 1826, d. 1908, Fellow of the Linnean Society.)

29. Bates, "Contributions...Heliconidae," p. 508.

30. Bates, "Contributions...Heliconidae," p. 510.

31. Bates, "Contributions...Heliconidae," p. 510.

32. Bates had observed the common color patterns found among groups of distinct species of unpalatable Heliconiidae. The phenomenon was analyzed by Fritz Müller and is now known as Müllerian mimicry. Neither Bates nor Wallace had a satisfactory explanation of the resemblances between unpalatable species. A discussion of Müllerian mimicry and its role in the Darwinian debate will be found in the chapter "Mimicry World-wide."

33. Bates, "Contributions...Heliconidae," p. 511.

34. Bates, "Contributions...Heliconidae," p. 512.

35. Bates, "Contributions...Heliconidae," p. 514.

36. Bates, "Contributions...Heliconidae," pp. 514-15.

37. Robert M. Stecher compiled all the known correspondence between Darwin and Bates. "The Darwin-Bates Letters, Correspondence between Two Nineteenth-Century Travellers and Naturalists," *Annals of Science 25*, Part I, March 1969, pp. 1-47; Part II, June 1969, pp. 95-115.

38. Stecher, "Darwin-Bates Letters," pp. 36-37. Letter, Darwin to Bates, November 20, 1862.

39. Stecher, "Darwin-Bates Letters," pp. 36-37. Letter, Darwin to Bates, November 20, 1862.

40. Many Darwin-Bates letters are also published in the biographical memoir published by Bates's neighbor and friend Edward Clodd. Edward Clodd, "Memoir" in *The Naturalist on the River Amazons. A Record of Adventures, Habits of Animals, Sketches of Brazilian and Indian Life, and Aspects of Nature under the Equator, During Eleven Years of Travel*, by Henry Walter Bates. With a Memoir of the Author (New York: D. Appleton, 1892) Letter, Bates to Darwin, November 24, 1862, pp. xlv-xlvi.

41. [Charles Darwin] [Review of] "Contributions to an Insect Fauna of the Amazon Valley," *Natural History Review 3*:219-224, 1863.

42. Darwin, *Nat. Hist. Rev. 3*, 1863, p. 223.

43. Darwin quoted this from Gray in his letter to Bates, March 4, 1863. Stecher, "Darwin-Bates Letters," p. 41. Asa Gray fully concurred. He wrote to Darwin stating, "I like and agree to your remark that in Bates' Geographical Varieties, etc., we get about as near to seeing a species made as we are ever likely to get." Jane Loring Gray, editor, *Letters of Asa Gray* (Boston, 1893). Gray to Darwin, July 7, 1863, Vol. II, p. 508.

44. Joseph Dalton Hooker (1817-1911) was a prominent botanist and Fellow and President of the Royal Society. E. Clodd, "Memoir," pp. xlvi-xlvii. Letter, Hooker to Bates, November 13, 1862.

45. It is fortunate that Edward Clodd published both the letters Bates received and those he sent. They have been lost. The Darwin-Bates letters are in the archives at Cambridge University Library.

46. Clodd, "Memoir," pp. xlviii-xlix. Letter, Bates to Hooker, January 11, 1862.

47. Clodd, "Memoir," pp. xlix-l. Letter, Hooker to Bates, January 13, 1862. Italics are Hooker's.

48. Clodd, "Memoir," pp. l-liii. Letter, Hooker to Bates, February 2, 1862.

49. Clodd, "Memoir," p. lii.

50. Clodd, "Memoir," p. liii. The analogy of centrifugal force in physics to the adaptive variations which radiate out from a center where they originated was a favorite of Hooker's. Here he used the analogy as a heuristic device to help Bates make the distinction between invisible causes and visible consequences of natural law. He wrote, "Finally, I have come to look upon the law of variation as I do upon gravitation; local circumstances may mask its effects, but upon *itself* they cannot act. The power that drives the stone to the ground is *the same* whether the stone permeates water, or air or vacuum, whether you let it fall straight, or throw it forwards, or even upwards. So a certain amount of centrifugal force of variation is distrubuted in certain proportions amongst the twelve peas in the pod, and except to arrest or retard the progress and amount of development of the individuals or their organs (I incline to think), local circumstances are powerless."

51. Clodd, "Memoir," p. liii. Letter, Hooker to Bates, February 2, 1862.

52. Clodd, "Memoir," p. liii. Hooker's earlier opinions may be found in *The Botany of the Antarctic Voyage of H. M. Discovery Ships Erebus and Terror, in the Years 1839-1843* (London, 1844-47). See the "Introductory Essay," to Volume II *Flora Nova-Zelandiae* (London, 1853) Chapter II, for Hooker's opinions on adaptation and permanent species.

53. Clodd, "Memoir," pp. liii-lv. Letter, Bates to Hooker, March 5, 1862.

54. Clodd, "Memoir," p. liv. The example of domestic ducks was only one of several which Bates gave in support of the idea that modifications induced by direct action are perpetuated by inheritance. Some of the other examples were the changes in wool texture in sheep transported into the tropics, the lop-eared condition of dogs which resulted from disuse, and blindness in cave species which was produced by disuse causing atrophy of muscles and nerves which is inherited by subsequent generations.

55. Clodd, "Memoir," p. lv.

56. Clodd, "Memoir," pp. lvii-lx. Letter, Bates to Hooker, March 21, 1862.

57. Clodd, "Memoir," p. lviii. Bates was inclined to accept the view "so prominently brought forth by Mr. Darwin that the reproductive functions are greatly disturbed when a wild species is brought under confinement or collection."

58. Clodd, "Memoir," p. lix. Bates wrote, "Acquired modification of the circulating fluids which affect the skin, its clothing and colors, the glandular system, etc., etc., seem more likely (than mutilations like circumcision and shaving) to be transmissible to descendents." Natural selection interacts, in Bates's view, with the direct effect of the environment as it does in adapting arctic animals to their surroundings. The temperature, Bates thought, "Could have something to do with the thick fur of arctic animals. But the white colors of the fur of those which live on the snow cannot be at all the effect of snow, but must be wholly due to natural selection."

59. F. Darwin, ed., _More Letters_ (New York: D. Appleton, 1903) 2 vols. I, p. 199. Letter, Darwin to Hooker, March 26, 1862.

60. Henry Walter Bates, _The Naturalist on the River Amazons. A Record of Adventures, Habits of Animals, Sketches of Brazilian and Indian Life, and Aspects of Nature under the Equator, During Eleven Years of Travel_ (London: John Murray, 1863) 2 volumes. Hereinafter referred to as _Naturalist_.

61. Darwin, _More Letters_, I, pp. 182-83. Letter, Darwin to Bates, April 4, 1861.

62. Stecher, "Darwin-Bates Letters," p. 34. Letter, Bates to Darwin, June 14, 1862.

63. Darwin wrote to Bates on October 15, 1862, "I want to hear a little news of you & your Book & how you & it go on." and again on October 18, 1862, "I am not _at_ all surprised at the turn which your Book has tekan. Everyone finds the same thing. It is dull work, but must be borne." pp. 34-36.

64. Stecher, "Darwin-Bates Letters," pp. 42-43. Darwin to Bates, April 18, 1863. Darwin went on to write, "You have spoken out boldly on Species; & boldness on this subject seems to get rarer & rarer."

65. Charles Darwin, [Review of] "The Naturalist on the Amazons," *Natural History Review* 3:385-389, 1863. Darwin did not want to leave the *Naturalist* to another's pen. The unsigned review emphasized the importance of the work for the "Darwinian theory" with carefully chosen quotations.

66. The second edition was also published by John Murray in 1864. A third, slightly abridged version was issued in 1873. It has been issued in several forms since. The most recent reprint, of the second edition, is by the University of California Press, 1962.

67. *Naturalist*, ed. 1, Ch. I, p. 19.

68. The following are some of the reviews of *The Naturalist on the River Amazons*. *The Athanaeum*, no. 1850, April 11, 1863, pp. 488-89. *Annals and Magazine of Natural History*, 12, 1863, pp. 591-96. *American Journal of Sciences and Arts* (Silliman's Journal) Nov. 1863, pp. 285-99. *Revue des Deux Mondes* 46, 1863, pp. 703-706. Finally, Bates's book was extensively noted by the Entomological Society President, F. P. Pascoe, in his address for 1864, pp. 240-245 of the *Transactions of the Entomological Society of London*.

69. *Naturalist*, p. 265. Some of the more important sections on natural history were pp. 206-209 on insect coloration, pp. 256-265 on the origin of species, and pp. 298-299 on mimicry in butterflies.

70. Stecher, "Darwin-Bates Letters," p. 34. Letter, Darwin to Bates, April 30, 1863.

71. A. R. Wallace, *My Life. A Record of Events and Opinions* (New York: Dodd, Mead & Co., 1905) 2 vols. Vol. I, p. 256.

72. Bates, *Naturalist*, pp. 206-207. Bates explained as simply as he could that conspicuous coloration may be equally as protective in its function as obscure coloration. In connection with the bright colors in mimicking insects he wrote: "I believe these imitations are of the same nature as those in which an insect or lizard is coloured and marked so as to resemble the soil, leaf, or bark on which it lives; the resemblance serving to conceal the creatures from the prying eyes of their enemies" (pp. 298-299). In discussing congenic tiger-beetles, *Tetrarcha pallipes* and *nocturna*, Bates had noted that the obscure beetle could hide better

and run faster than the brighter tiger-beetle. It had the protection of a disagreeable odor which made it less attractive as prey. Bates concluded, "Thus we see that the fact of some species not exhibiting the same adaptation of colors to dwelling-places as the companion species does not throw doubt on the explanation [natural selection] given of the adaptation, but is rather confirmatory of it" (p. 208).

73. Bates, Naturalist (first ed.), pp. 345-346 of Volume II. (second ed.) pp. 412-413. Bates is not alone in his self-conscious and self-defensive apologetic for entomology. The president's addresses in the Proceedings of the Entomological Society of London contain many similar statements that stress the utility of their science. Particular emphasis is made that even the study of the butterflies has real usefulness.

74. John Edward Gray (1800-1875) was one of the most vitriolic anti-Darwinians, which may have been partially due to the fact that most Darwinians were both heretics and amateurs. Darwin also recalled having been attacked by him. Darwin recalled Gray having said, "You have just reproduced Lamarck's doctrine, and nothing else, and here Lyell and others have been attacking him for twenty years, and because you (with a sneer and laugh) say the very same thing, they are all coming round; it is the most ridiculous inconsistency, &c., &c." Darwin, Life and Letters, II, pp. 242-43. A new book has just been published which discusses J. E. Gray. Albert E. Gunther, A Century of Zoology at the British Museum Through the Lives of Two Keepers, 1815-1914 (New York: Science History Publications, 1975).

75. Clodd, "Memoir," p. lxvi. Letter, Hooker to Bates, May 13, 1863.

76. Clodd, "Memoir," p. xlvi. Letter, Bates to Hooker, May 12, 1863. Bates applied for an opening in the British Museum but knewthat he stood little chance of an appointment with the opposition of Richard Owen and other anti-Darwinians. The position was given to a poet who knew little or no entomology, Arthur O'Shaughnessy. George Woodcock writes, "He [Bates] was a scientific heretic and the Museum authorities preferred an ignoramus to a Darwinian." Henry Walter Bates, Naturalist of the Amazons 1969, p. 250.

77. William Wilson Saunders (1809-1879) F. R. S. and twice president of the Entomological Society of London. He was the owner of a very large collection. Saunders had agreed that the papers would be issued under Bates's name, and would

be illustrated by Westwood, whose skill was highly respected. Bates worked on Mantidae as well as Phasmidae in Saunders's collection.

78. Clodd, "Memoir," p. lxv. Letter, Bates to Hooker, May 12, 1863.

79. Clodd, "Memoir," p. lxx. Letter, Bates to Darwin, September 29, 1863. The "monograph" was on the Mantidae group of Orthoptera to which the Phasmidae also belong.

80. Henry Walter Bates, "Descriptions of Fifty-Two New Species of Phasmidae from the Collection of Mr. W. Wilson Saunders, with Remarks on the Family," Transactions of the Linnean Society of London 25:321-359, 1865-66. Read June 15, 1865.

81. General camouflage is called "mimesis" rather than mimicry by some modern writers, for example Wolfgang Wickler uses that term in Mimicry in Plants and Animals, translated by R. D. Martin (New York: World University Library, 1968). It is a valid distinction, but the term "mimicry" was used by most nineteenth-century writers for both general camouflage to inorganic or plant substrates as well as for mimicry in which one insect resembles another for its own protection.

82. Bates, "Descriptions...Phasmidae," pp. 323-24. Adolph Gerstaecker (1807-1892) believed, apparently, that the classification based on leg armatures was more "natural" than the wing classification system. Bates wrote: "The wording of the objections to the alary system by the Berlin Professor, as in all such cases, implies that there exist in the nature of the thing the materials for a neatly definable classification of the group, which the Monographer had failed to discover. But I believe such materials do not in this case exist."

83. Bates, "Descriptions...Phasmidae," p. 324.

84. Bates, "Descriptions...Phasmidae," p. 323. Natural selection operated in the evolution of species in both the heterogeneous and homogeneous type. In some forms the adaptations are more limited in the number of body parts modified, that allows the taxonomist to use the unaffected parts for his classification, but in cases like Phasmidae all corporeal parts became variable, "Natural selection having from the first favoured the species which offered

variation in these parts... A group like the Phasmidae, therefore, ought, instead of perplexing us, to be regarded with high interest, as showing that the existence of generic characters is not a universal rule, and as affording a probable explanation why it is not" (p. 323).

85. John Obadiah Westwood (1805-1893), Catalogue of Orthopterous Insects in the Collection of the British Museum. Part I Phasmidae (London, 1859). The other authorities were Hermann Burmeister (1807-1892), an honorary fellow of the Entomological Society of London, Jules Lichtenstein (1816-1886), John Edward Gray (1800-1892), and Willem DeHaan (1801-1855).

86. Bates, "Descriptions...Phasmidae," p. 326.

87. Bates, "Descriptions...Phasmidae," p. 326.

88. Darwin described his way of working in his autobiography. "I worked on true Baconian principles, and without any theory collected facts on a wholesale scale." Nora Barlow, editor. The Autobiography of Charles Darwin, 1809-1882. (New York: W. W. Norton, 1969), p. 119. The terms "Baconian science" and "inductive method" were often very nearly synonymns for "good scientific practice." Bates was also fond of the term "inductive science." Another installment of "Contributions to an Insect Fauna of the Amazon Valley. Lepidoptera: Papilionidae," Journal of Entomology, Descriptive and Geographical. 1, no. 4, December 1861, pp. 218-245, concluded with these remarks: "Discussions upon the derivation and relationship of local varieties are of great scientific interest; but they only reach their true value when they are applied also to the apparently more distinct forms. It is not, I conceive, according to the spirit of inductive science that the method which is applied to prove the natural derivation of a semi-distinct form should be relinquished on the incorporation of the variety with the species" (p. 223).

89. Bates, Naturalist, pp. 260-261.

90. Stecher, "Darwin-Bates Letters," p. 11. Letter, Bates to Darwin, March 28, 1861.

91. Morse Peckham, editor. The Origin of Species, by Charles Darwin. A Variorum Text. (Philadelphia: University of Pennsylvania Press, 1959), pp. 663-670.

92. Darwin, The Origin of Species, 6th edition (New York: Modern Library, n.d.), pp. 326-327. Jean Baptiste

Pierre Antoine de Monet de Lamarck (1744-1829); William Sharp MacLeay (1792-1865).

93. Darwin, Origin, 6th ed., p. 328.

94. Darwin, Origin, 6th ed., p. 328.

95. The presence of analogous organs in disparate groups was well known and had been perceived as a difficulty for evolutionists. It had been a phenomenon well adapted to illustrate the argument from design. Darwin was consequently very eager to use a case of analogous resemblance to illustrate natural selection. The analogies of variation and adaptation in different taxonomic categories had been used by MacLeay and other Quinarians as the basis for classificatory schemes. Darwin wanted there to be no doubt that the analogies could have been produced by the preservation of variations under natural selection's operation. Darwin took great care to discuss that mimicry could have been developed gradually and that it was not created.

96. Darwin, Origin, 6th edition, p. 330.

97. Peckham, Variorum, p. 666.

98. Peckham, Variorum, p. 666.

99. Charles Darwin, The Descent of Man and Selection in Relation to Sex (London: John Murray, 1872). The edition used here is the Modern Library (New York, n.d.). Mimicry is discussed in Chapter XL, pp. 666-668.

100. Roland Trimen, "On some Remarkable Mimetic Analogies Among African Butterflies," Transactions of the Linnean Society of London, 26, 1870, p. 497; and A. R. Wallace, "On the phenomena of Variation and Geographical Distribution, as Illustrated by the Papilionidae of the Malayan Region," Transactions of the Linnean Society of London 25:1-72, 1865; and "Mimicry and Other Protective Resemblances among Animals," Westminster Review, July 1867.

101. Stecher, "Darwin-Bates Letters," p. 112. Letter, Darwin to Bates, April 22, 1868.

102. Stecher, "Darwin-Bates Letters," pp. 116-17. Letter, Bates to Darwin, January 17, 1870.

103. Stecher, "Darwin-Bates Letters," p. 117.

104. Michael T. Ghiselin, The Triumph of the Darwinian Method (Berkeley: University of California, 1969). The term "Darwinian" has had a strange career of varied uses. It has meant everything from "transformist". to "neo-Lamarckian," and "neo-Darwinian," which Darwin never was. It is very important to credit the other Darwinians for their theoretical and factual contributions of Darwinian data.

105. Woodcock, H. W. Bates, Naturalist...Amazons, p. 259.

106. Some of the biographical sketches of Bates are the following: F. D. Godman, Transactions of the Entomological Society of London (Proceedings) 1892, i-iv; Robert McLachlan, Entomologists' Monthly Magazine 28:83-85, 1892; Royal Geographical Society Proceedings 14:245-257, 1892; David Sharp, The Entomologist 25:77-80, 1892; Alfred Russel Wallace, Nature 45:398-399; H. Lewis McKinney, Dictionary of Scientific Biography.

107. Stecher, "Darwin-Bates Letters," p. 13. Letter, Bates to Darwin, March 28, 1861.

Chapter III

MIMICRY WORLD-WIDE

Bates's demonstrations of "palpably intentional" resemblances between edible and distasteful butterflies seemed to Darwinians to have all the qualities of elegance and beauty typical of a proof of natural selection. Bates's book, *The Naturalist on the River Amazons* (1863), brought the subject of mimicry and its relationship to the process of natural selection into the parlors of the readers of adventure stories as well as into the libraries of scholars.[1] It was a good sequel to the paper on mimicry published in the *Proceedings of the Linnean Society* (1862) where mimicry was presented as a striking zoogeographical phenomenon.[2] In the *Naturalist*, Bates reported his discovery of mimicry in butterflies as a personal adventure, like his first view of the tropical forest and his encounter with giant alligators and with tarantulas large enough to capture finches. Although mimicry in butterflies was presented as another tropical adventure, he did not spare his readers the knowledge that the author of the *Naturalist* was a committed Darwinian.[3] He made his position clear: that no other causes besides natural ones were needed to explain mimicry. To him, natural selection alone was sufficient to initiate and perfect the remarkable likeness between structurally different groups of insects. Even in the Naturalist, Bates tried to be sure the readers did not associate his view of change in animals with Lamarck's.[4] Similarity between models and mimics did not arise because the insects were exposed to similar physical conditions. The conformity Bates could report of observable data to Darwinian principles was just the kind of findings needed if the theory of natural selection was to gain credibility among observation-oriented naturalists of the period. Bates's work, both in scientific and in popular literature was buttressed by publications of travelers in other parts of the world in the 1860's and 1870's.

In the early phase of the Darwinian controversy, mimicry was the single best exemplar of analogical resemblances having been developed through natural selection. Darwin acknowledged this in the fourth edition of the Origin of Species (1866), where he wrote:

> The most remarkable case of analogical resemblance ever recorded, though not dependent on adaptation to similar conditions of life, is that given by Mr. Bates with respect to certain butterflies in the Amazonian region closely mimicking other kinds...so that here we have an excellent illustration of the principle of natural selection.[5]

By the date of this edition Darwin could mention Wallace's corroborative findings in the Malay Archipelago and by the sixth edition of 1872 he also could mention Roland Trimen's discovery in South Africa.[6] Darwin was hopeful that the explanation of mimicry by natural selection would be convincing to naturalists that natural selection was effective in changing organisms. He knew that the explorers would, if they could, report new cases of mimicry. It seemed quite likely that more data on the subject of mimicry would improve the chances of acceptance of the Darwinian theory.

The naturalists who found mimetic butterflies in distant faunal provinces from the original place Bates found them in South America were all Darwinians. When the papers were published discussing mimicry in South-East Asia, Africa, Central America and North America, the phenomena were introduced as examples of the efficacy of natural selection. Benjamin Walsh and Charles Riley, the finders of mimicry between the American Monarch and Viceroy butterflies, discussed the subject as an example of evolution and natural selection.[7] Bates's popular book of natural history was succeeded by Thomas Best's *The Naturalist in Nicaragua*, which also presented mimicry in Darwinian terms accessible to the readers of adventure literature.[8]

Some of the first cases of mimicry extending the Batesian concept to other faunal provinces of the world have remained interesting and controversial among entomologists. Wallace's discovery of mimetic polymorphism in *Papilio polytes* in the Malay Archipelago and Roland Trimen's discovery of complex mimetic relationships in *Papilio dardanus* in Africa are still often named in modern textbooks.[9] Some of the most intense debate about evolutionary problems focuses on a few species rather than more diffusely on the theory itself.

The subject of mimcry and Darwinian explanation changed somewhat after 1879 when Fritz Müller, a German defender of Darwin, found a new kind of mimicry in South American butterflies.[10] This time, all the species were protected from predation by distastefulness and were similarly colored. Müller explained this type of mutual mimicry among unpalatable species as also illustrating the action of natural selection.[11]

The authors of papers extending Batesian mimicry geographically and introducing Müllerian mimicry were offering data supporting the Darwinian position. However, many naturalists who opposed Darwinian evolution were quick to respond to the claims that mimicry proved or supported the action of natural selection. In many cases they reinterpreted findings in the same species the Darwinians used. Did additional data help to support the Darwinian position as Bates and Darwin had expected it would? In following the arguments for and against the Darwinian interpretation of mimicry in a few species, it is possible to gain some sense about the relationship of observations to theory.

Wallace, Papilionidae, and *Papilio Polytes*

Bates's work on mimicry did not have to stand alone for very long. Only two years later, Alfred Russel Wallace read a paper on mimicry in a new geographical area at the Linnean Society on March 17, 1864. Once again the word "mimicry" did not appear in the title. According to tradition, Wallace kept his title simple and descriptive, "On the Phenomena of Variation and Geographical Distribution as Illustrated by the Papilionidae of the Malayan Region."[12] Wallace, like Bates, was interested in establishing the basis for a larger project on geographical distribution; and the Papilionidae was a family rich enough in species, and yet limited enough in range, to allow him to observe island-to-island endemism. These local changes supported his view that the Malay Archipelago was made up of two zoogeographically distinct regions, the Indo-Malayan and the Austro-Malayan.

It is one of Wallace's most inductive papers and it was written so that the newly discovered species could contribute toward revising general conclusions on the nature of species, divergence, zoogeographical provinces. It also explained in Darwinian language the meaning of mimetic analogies in the Papilionidae. Wallace's paper on the Malayan Papilionidae provides us with capsule summaries of many of the topics he expanded with the labors of succeeding years.

The argument presented in the essay on "Malayan Papilionidae" touched on questions in both taxonomy and zoogeography. It presented an account of the zoogeography of the family and it sorted out the different kinds of variation found in the family. Wallace proceeded then to account for the preservation of adaptive variation and to argue for gradual speciation on what he liked to call "Mr. Darwin's principles." In order to establish a basis on which to discuss the different degrees of variation he had found in Papilionidae and for the purpose of relating variation to evolution, Wallace recognized six kinds of variation. Wallace's list of types of variation was made up of the following phenomena: "1st, simple variability; 2nd, polymorphism; 3rd, local forms; 4th, coexisting varieties; 5th, races or subspecies; and 6th, true species."[13]

The discussion of the distinguishable phenomena of variation presented a case similar to the one Bates had made in his paper in 1863 on the South American Heliconiidae. Both naturalists expressed a sense of hopelessness about identifying unequivocal characteristics for species discrimination. The longer and harder one worked with zoogeographical problems, the harder it became to find sharp distinctions on which to base taxonomic distinctions. Wallace believed that a new vision of zoological research was needed rather than mere refinement of technique or additional information. The vision or sense of need for a new philosophy of zoology grew from his attention to a different sector of zoology from that which dominated British natural history. Wallace's work on zoogeography, particularly his study of mimetic polymorphism in butterflies, made him think that most naturalists were looking at the wrong level of integration. In attempting to define species, they neglected other degrees and kinds of variation. The more one looks at varieties and races, the more one becomes convinced of the instability of the species. It is from the instability of variation, in categories one through five, that it is possible to learn the true nature of category six, the true species. Wallace wrote:

> It is for such reasons that naturalists now look upon the study of varieties as more important than that of well-fixed species. It is in the former that we see nature still at work, in the very act of producing those wonderful modifications of form, that endless variety of colour, and that complicated harmony of relations, which gratify every sense and give occupation to every faculty of the true lover of nature.[14]

The study of the gradations of variation beneath the species level reveals more precisely than studying the higher taxa can that animals change in time in relation to their environmental conditions. The study of the order Lepidoptera was particularly well-suited to Wallace's view, since the Lepidoptera are ecologically quite similar throughout the group.

All Lepidoptera are vegetarians. As larvae they all eat plants; the adult (imago) forms feed by means of sucking mouth parts. One would expect on the basis of this ecological uniformity that the population of Lepidoptera would be roughly equivalent to that of another insect order which pursues a single mode of life. But what one finds is that the order of butterflies and moths is numerous beyond the naturalist's expectation. The abundance of butterflies proved to Wallace's satisfaction that special modifications and adaptations permitted the great proliferation of Lepidoptera within the single ecological mode of life. The butterflies were a group prone to vary and, also, involved in complex relationships with animals which prey on butterflies.

Both Wallace and Bates emphasized the importance of predation on the preservation of adaptive variations. Survival in the struggle for life was, after all, a process basic to evolutionary change in species. The relationship between variation and predation would be studied in the Malayan Papilionidae by careful study of their external forms and color patterns. Wallace held that insectivorous animals, especially birds, were the agents of natural selection with respect to insects. The real challenge then lay in discovering the nature of the adaptations observed in butterflies to escape predation and their connection to Darwinian principles.

Mimicry as a means of escaping predation provided Wallace with a body of data relating observable characteristics of color and form with natural selection. Furthermore, the Malayan Papilionidae presented some amazing cases of mimetic butterflies in which the female of the species alone was a mimic. Wallace found a case of remarkable female polymorphism in the species *Papilio polytes*, which has three different female forms.[15] (See Plate III.) One of the female forms resembles the male, while the other two forms mimic two different species of Papilionidae which are distasteful to predators. Each of these three female morphs had been regarded as a separate species: *P. polytes*, *P. romulus*, and *P. pammon*.[16] The knot of species and forms which resembled members of other species presented very great difficulties. Many of the polymorphic species were related to the species they mimicked by affinity as well as by analogy in color. It was difficult enough to discover that three morphological species are members of a single biological species, but this was compounded by the separate matter of the taxonomy of the whole group. *Papilio polytes* was a species which illustrated both polymorphism and sex-limitation of mimicry. It demonstrated the lengths to which a species could go in response to natural selection. In respect to the sexual dimorphism and to the mimetic polymorphism in _P. polytes_, Wallace felt that he had good data related to gradual evolutionary adaptation of species.

The zoogeographical distributions of the _Papilio_ species gave him clues relevant to competitive advantage of one morph over another. He found that one morph was dominant in Java and another on Sumatra. He was unable to specify the conditions which made the forms differentially successful, however. Nevertheless, the key issue was that natural selection suffices to produce extremely marked cases of sexual dimorphism, if that is the better strategy for the species. In other cases natural selection may produce mimetic resemblances involving one or both sexes of a species, if that is the better means for the preservation of the species. Wallace stated:

> We have an indication of how dimorphism may be produced; for let the extreme Philipine forms be better suited to their new conditions of existence than the intermediate connecting links, and the latter will gradually die out, leaving two distinct forms of the same insect, each adapted to some special conditions.[17]

Although he could not specify the details conferring advantage on one morph over another, the most important factors in his discussion are the pressure of predators on the species and the competition of other insect species competing for the same ecological niches and habitats. For Wallace, ecological conditions and the adaptations in species in the struggle for life almost always took precedence over the struggle for reproduction. The differences between the male and female butterflies were due, more often than not, he thought, to differences in the habits of the two sexes. The fact that females were often mimics or were better adapted to concealment indicated to Wallace the predominant action of natural selection over sexual selection. Sexual selection had a very subordinate role in the development of sexual dimorphism or in the development of female polymorphism.

Sexual dimorphism in color was basically similar to polymorphism, including mimetic polymorphism. It was produced by natural selection carried out by predatory birds. Each morph had been perfected independently of the other forms of the same species. Hybrids were almost never found between the three female forms.[19] Wallace could not solve the genetic problems, but went on the assumption that natural selection had produced the hereditary integrity of the different forms. To illustrate the discussion of polymorphism, Wallace related an enigmatic little story which probably did little to assuage the doubts of the detractors of natural selection.

> The phenomena of dimorphism and polymorphism may be well illustrated by supposing that a blue-eyed, flaxen-haired saxon man had two wives, one a black-haired, red-skinned Indian squaw, the other a woolly-headed, sooty-skinned negress--and that instead of the children being mulattoes of brown or dusky

> tints, mingling the separate characgeristics of their parents in varying degrees, all the boys should be pure Saxon boys like their father, while the girls should altogether resemble their mothers. This would be thought a sufficiently wonderful fact; yet the phenomena here brought forward as existing in the insect-world are still more extraordinary; for each mother is capable not only of producing male offspring like the father, and female like herself, but also of producing other females exactly like her fellow-wife, and altogether differing from herself. If an island could be stocked with a colony of human beings having similar physiological idiosyncrasies with *Papilio Pammon* or *Papilio Ormenus*, we should see white men living with yellow, red, and black women, and their offspring always reproducing the same types; so that at the end of many generations, the men would remain pure white, and the women of the same well-marked races as at the commencement.[20]

Although Wallace wanted his parable of the mixed marriage to underscore the point that there is a very great difference between the female morphs and between male and female forms, he did not want to convey the impression that only a flying leap of variation could produce polymorphism. He wanted to show, rather, that the great differences between morphs had been produced by a very powerful cause in natural selection. Wallace knew that many readers of his paper would object that his study of mimetic polymorphism was only a rare and unimportant phenomenon which did not bear the general significance he wanted to attach to it. Wallace wrote, attempting to circumvent that objection:

> I reply that a true law embraces all apparent

> exception, and that to the great laws of nature there are no real exception--and that what appear to be such are equally results of law, and are often (perhaps indeed always) those very results which are most important as revealing the true nature and action of the law.[21]

The great law of evolution by natural selection includes the unusual cases in which a single species like _Papilio polytes_ has a single male form but three different female forms, only one of which is like the male, two resembling closely other butterfly species. These unusually striking phenomena reflect nature's most general laws: the differential survival of individuals having adaptive variations.

The mimetic polymorphism of _P. polytes_, and other species, so strongly marked in Wallace's time, had been slowly perfected from an ancestral stock in which female and male forms were probably similar. Unequal susceptibility to predation in the two sexes resulting from different ecological relations and sex-limited variation had resulted over many centuries in the remarkable di- and poly-morphism found in some Malayan Papilionidae. Polymorphism within a single species provided Wallace with an excellent example of how natural selection can cause the separate adaptation of the sexes of a species.[22] Polymorphism served Wallace best to show how different forms had been produced over time in response to the natural selection of adaptive variations in specific geographical circumstances. The phenomena of mimetic polymorphism, however, were not so good for showing the process of speciation by small steps. Local forms and subspecies were better categories of variation for showing sequential steps of speciation. In spite of a fine series of examples of mimetic polymorphism and of geographical species phenomena in the Papilionidae, Wallace still was unable to present an argument tight enough to satisfy readers hostile to Darwinian principles. Wallace's discussion required that the reader grant that mimetic and near-mimetic individuals have a differentially better chance of surviving to reproduce than individuals not having those variations. His position entailed the belief that hereditary transmission of individual variations takes place, and last, that further hereditary variations tending to perfect mimetic resemblances be forthcoming. Of course, besides these requirements tending toward the development of mimicry, the mating relationship between a constant male form and different morphs of females would have to be maintained.

For Wallace, mimicry in the female sex alone and especially mimetic polymorphism involving the female only had important bearing on the whole question of whether sexual selection or natural selection was the more important factor behind sexual dimorphism. His studies of Papilionidae species like _P. polytes_ seemed to Wallace to confirm the position that insect colors are generally protective adaptations preserved by natural selection. Mimicry in the female sex was like protective coloring in the female sex. Either strategy produced the same result of protecting the individual in the struggle for life. Cases in which male and female differ in color must be judged by obtaining thorough knowledge of the habits of life of each sex.[23] When females are protectively colored, either by cryptic colors or by mimetic color patterns, more than the males of the species, Wallace believed that natural selection had produced the observed results over a long evolutionary period of differential survival.

> The reason why the females are more subject to this kind of modification than the males is, probably, their exposure to attack while in the act of depositing their eggs upon leaves, render it especially advantageous for them to have some additional protection. This they at once obtain by acquiring a resemblance to other species which, from whatever cause, enjoy a comparative immunity from persecution.[24]

Adaptation, for Wallace, was a natural concomitant of spontaneous variation in individuals in the presence of predators acting as the agents of natural selection. Darwin's view that sexual selection produced most of the cases of secondary sex differences in animals, including insects, seemed to Wallace to have been vitiated by his study in Malayan Papilionidae. Even if one were prepared to admit that many of the differences found between male and female butterflies were due to mating preferences rather than to protective adaptation, how could it explain the presence of three different female forms in one species? In Wallace's paper, natural selection alone could account for the entire body of data he had compiled from the Papilionidae.

In Wallace's paper, considerations of mimetic phenomena were linked with geographical findings. The discoveries supported the conclusion that mimicry had a protective result for the individuals having it. Conspicuous coloring could be as protective as dull, cryptic colors, provided that it was deployed in a manner that deceived a predator into mistaking a palatable insect for an unpalatable one. The cases of butterflies which had more than one mimic form in a single species were particularly interesting to Wallace since they indicated the lengths to which natural selection could go within a single species. The combination of phenomena of variation in the Papilionidae seemed to jointly deny the stability of species called for by the theory of special creation. Wallace asked in a series of rhetorical questions whether explanation predicated on the theory of special creation could satisfactorily explain island-to-island changes in the butterfly fauna.[25] It seemed to him inadequate to expalin the phenomena of the Papilionidae: they were too variable, too local, and often too imperfect in their mimicry of models or in other protective adaptations. The Malayan Papilionidae required an understanding of species that accounted for divergence of species, geographical shifts in species, island-to-island variations, and imperfect degrees of adaptation. For him, natural selection was the required principle.

Papilio Polytes and Discontinuous Variation

Although mimetic polymorphism seemed to Alfred Russel Wallace the paradigmatic example of action of natural selection, this same phenomena was selected by several early geneticists as an example of evolution by saltation. In fact, the species Wallace had studied very closely, *Papilio polytes* was the same one used in an early paper on discontinuous variation. J. C. F. Fryer, a student of the geneticist Reginald Crundell Punnett, does not mention Wallace at all in his paper of 1913.[26] As was typical in the early years of the twentieth century for students of genetics, Fryer doubted the importance of natural selection in evolution. So far as he could tell, in respect to *Papilio polytes*, natural selection was not active. The criterion he had set to measure whether or not natural selection was operating was the ratio of the non-mimetic female morph (*cyrus*) of *Papilio polytes* with respect to the other two mimetic forms. Since Fryer had been able to determine that the numbers of non-mimetic forms (*cyrus*) were not

EXPLANATION OF PLATE III

Figures:

1. *Papilio polytes* male, non-mimetic.
2. *Papilio polytes* female, var. *cyrus* (M form), resembling the male, Fig. 1.
3. *Papilio polytes* female, var. *polytes* (A form), mimic of *P. aristolochiae*, Fig. 5.
4. *Papilio polytes* female, var. *romulus* (H form), mimic of *P. hector*, Fig. 6.

1a-6a. Undersurfaces of the hind wings of each specimen.

Specimens figured on the plate were from Ceylon.

PLATE III

MIMETIC POLYMORPHISM IN PAPILIO POLYTES

(From R. C. Punnett, Mimicry in Butterflies, 1915, Plate V).

waning in respect to the other forms, he believed the view was warranted that no effect of natural selection was present, at least no effect that could be measured. He reasoned that since mimicry was an adaptive trait conferring protection on unprotected butterflies, that trait should spread throughout the population until no more non-mimetic *cyrus* females were left.

Fryer's work with *Papilio polytes* was an attempt to add rigor to the study of the units of inheritance. In order to do this he tried to characterize the different morphs of the species with letters standing for Mendelian characters.[27] His hope was that by breeding butterflies of known pedigree it might be possible to work out precisely the ratios of different forms of offspring born to different parental forms. The situation in *P. polytes* was quite complex. Not only were characters for dominance and recessive important, but sex factors also had to be considered. Besides dominance and sex, an inhibiting factor or gene was active inhibiting the formation of mimetic males in the species. Fryer, like many other biologists in 1915, was looking to genetics for answers to species characters, rather than to natural selection or to evolutionary theory. It seemed to them that the selection played little or no role in species they were studying, like *P. polytes*. It seemed all the more certain that natural selection was inactive because Fryer's data conformed to G. H. Hardy's formula for stability in Mendelian ratios in a population.[28]

Fryer's paper on Papilio polytes was quite quickly put to a more polemical use in the cause of advancing the saltation theory of evolution. His teacher, Reginald Punnett, as all early Mendelians, was an enthusiastic advocate of evolution having occurred by large mutations.[29] It seemed to him that natural selection had been much over-rated as an agency in the origin of species. Punnett went so far as to suggest that natural selection as it was discussed by the Darwinians was like the ancient and discredited natural theology of former times. Punnett stated that the Darwinians, just like the natural theologians, were forced to contrive uses for the structures found in organisms. He argued:

> The whole matter is too often approached in much the same spirit as that in which John Ray approached it two centuries ago, except that the Omnipotency of the Deity is replaced by the Omnipotence of Natural Selection. The vital point, which is whether Natural Selection does offer a satisfactory explanation of the living world, is too frequently lost sight of. Whether we are bound or not to interpret all the phenomena of life in terms of natural selection touches the basis of modern philosophy. It is for the biologist to attempt to find an answer, and there are few more profitable lines of attack than a critical examination of the facts of adaptation. Though "mimicry" is but a small corner in this vast field of inquiry, it is a peculiarly favourable one owing to the great interest which it has excited for many years and the consequently considerable store of facts that has been accumulated. If then we would attempt to

> settle this most weighty point in philosophy there is nothing to which we can appeal with more confidence than to the butterfly.[30]

Nothing could have been more irksome to the selection theorists than to have been accused of retaining the spirit of natural theology. They had wielded that very accusation in their attacks on Lamarckians and orthogenesists. The geneticists were now the ones writing with the confidence that their view of biology was to be the organizing principle of the future.

Punnett's general assault on evolutionary theory of the "old school" was followed by point-by-point rebuttal of Wallace's view of mimicry and by two chapters "On the Case of *Papilio polytes*."[31] To Wallace's first condition of mimicry (1) "That all the imitative species occur in the same area and occupy the very same station as the imitated" Punnett named cases in which models and mimics were not observed to fly together.[32] There were exceptions also to two other of Wallace's laws of mimicry: (2)"that the imitators are always the more defenceless" and (3) "that the imitators are always less numerous in individuals."[33] Punnett was quick and cavalier in dismissing the criteria of mimicry that Wallace had worked out, including the condition that the mimic had "departed from the bulk of its allies."[34] In ten very small pages of ten paragraphs, Punnett abruptly concluded, "Clearly there is a *prima* facie case for doubting whether we must of necessity ascribe all resemblance of the kind to natural selection."[35] By trying to show that Wallace's laws of mimicry were constantly disobeyed, he hoped that Wallace's explanation of the evolution of mimetic resemblances would fall along with the laws.

Perhaps Wallace had overreached a little in framing the laws of mimicry in absolute terms without qualifying words. He wrote in _Darwinism_ (1889), "The imitative species occur in the same area and occupy the very same station as the imitated."[36] He had strengthened his claim since he had first enunciated criteria for mimicry in the _Contributions to the Theory of Natural Selection_ (1871), where he had written, "The first law is, that in an _overwhelming majority_ of cases of mimicry, the animals (or the groups) which resemble each other inhabit the same country, the same district, and _in most cases_ are to be found on the very same spot."[37] When the _Contributions_ were published, Wallace discussed mimicry as part of an argument that they had previously been inadequately treating as merely being "curious analogies." In the discussion in _Darwinism_, some years later, the context in which mimicry was placed was different. In the later work, the five laws of mimicry were stated in stronger terms, showing that the phenomena had definable characteristics. The laws of mimicry were stated as general criteria which are best understood as worked on in species by natural selection. Darwinism presents Wallace's evidence of mimicry that is never produced by the agencies proposed by the detractors of Darwinian theory, like "like condition," "accidental resemblances," and "reversion to ancestral types."[38]

Between the publication of Wallace's paper on the "Malayan Papilionidae" in 1865 and the publications of Fryer and Punnett in 1913 and 1915, major changes had occurred in biological study. Scientific philosophy was changing and especially with the rise of experimentation, the level of integration given most attention by biologists was shifting. As a naturalist, Wallace had given most of his attention to the organism and to the populations and species of organisms. He wrote in 1865:

> I must here express my own belief that none of these phenomena, however apparently isolated or insignificant, can ever stand alone--that not the wing of a butterfly can change in form, or vary in colour, except in harmony with, and as part of, the grand march of nature.[39]

That statement stands very near to being Wallace's credo. For him the charm of zoological study lay in the solving of puzzles in terms of natural causes and laws. Whenever the coloration of animals was at issue, the explanation Wallace favored as having greatest force was the Principle of Utility enforced by the process of Natural Selection. Mimicry was a special case of protective coloration in which animals accomplish through conspicuous markings the same goal of survival accomplished in other cases by cryptic, obscure markings. Polymorphism was one of the types of variation which when coupled with mimicry served a species by making it possible to expand both its population and its range. In his early work, the topic of mimicry was subordinated to the larger topic of zoogeographiy on which Wallace was working. Attention to geographical phenomena is very different from attention to the hereditary transfer of particular traits in terms of what is accepted as evidence in support of a theory.

Wallace opened his article in the Westminster Review, "Mimicry, and other Protective Resemblances Among Animals," with a philosophical introduction.[40] He began by addressing the subject of the differences between true theories and false ones, which differ in their capacities to interpret phenomena which had stood as anomalies to older theories. The false theories which Wallace singled out to demonstrate this principle were the polarity theory of Edward Forbes and the Quinary classification system of MacLeay and Swainson.[41] These theories had had short lives because of the masses of data they failed to incorporate. He wrote that natural selection theory was earning rapid acceptance among scientists because it was able to explain both old and new findings. "New faces, new problems, new difficulties, as they arise are accepted, solved, or removed by this theory; and its principles are illustrated by the progress and conclusions of every well-established branch of human knowledge."[42] Wallace's sangine attitude was based on his work as a zoogeographer.

The geological component of Wallace's zoogeography work indicated that the earth had undergone slow, uniform changes. Land connections had been made and broken. Divergence of the biota had been occurring over a greatly protracted period of time. Wallace was never convinced by the arguments of the physicists that the age of the earth was short. For Wallace, there was plenty of time for slow evolution to occur by descent with modification. Wallace could find no reason not to hold that the diversification of animal coloration had occurred gradually in response to changes in distribution and in response to predators. For most animal coloration patterns protection was their purpose, and utility was the criterion by which the purpose was defined. The many ways of eluding predators had gradually evolved and been perfected as the predators themselves had changed in the struggle for life. The case of Papilio polytes fitted with evolutionary views on geographical grounds. Results must fit together from studies of variation and predator behaviour as well as from zoogeography if the Darwinian theory was to have the all-encompassing explanatory power Wallace asserted that it had.

In spite of all Wallace could do, the burden of proof remained on the shoulders of the Darwinians. After the rediscovery of Mendel's paper at the turn of the twentieth century, interest shifed to the experimental study of genetics and left the organism-level zoologists, whose guiding principle was Darwinian evolutionary theory, even further isolated from the mainstream of zoological thought. The period from 1910 to 1920 was a decade of escalating confidence in the power of experimental breeding studies to reveal the true story of evolution. The earliest Mendelian geneticists were confident that the steps evolution had taken could be recaptured by back-crossing experiments to reveal the genotype of parental forms.

It seemed to men like Punnett and Fryer that their experiments with *Papilio polytes* showed that since the earlier steps could not be retraced by crossing various morphs with the male of the species, that there had been no steps from the non-mimetic to the mimetic conditions. The absence of intermediates between the morphs of *P. polytes* for the geneticist, amounted to the disproof of gradual evolution by natural selection. Fryer's work with the *P. polytes* forms was corroborated by the geneticists working with plants.[43] Punnett wrote in 1915:

> From experiments with cultivated plants such as primulas and sweetpeas, we have learnt that this discontinuous form of inheritance which occurs in *P. polytes* is the regular thing. Moreover, we have plenty of historical evidence that the new character which behaves in this way is one that has arisen suddenly without the formation of intermediate steps.[44]

Punnett was confident that the analogy between flowering plants and Lepidoptera was reliable. Since sweet peas had changed suddenly and spontaneously without small intermediate gradations and since *P. polytes* responded to breeding experiments in much the same way, it seemed to him that the analogy was good. He wrote of **the morphs,** "There is every reason to suppose that they arose in the same way, that is to say, as sudden sports or mutations and not by the gradual accumulation of slight differences."[45] Natural selection may be rather certainly eliminated, Punnett asserted, from having any important role in the origin of mimicry. Furthermore, the part natural selection plays in conserving the new adaptations provided by mutations may be studied by careful numerical investigations. If new mimetic forms are favored in the struggle for life over the older non-mimetic forms, there should be a differential increase in the numbers of mimetic to non-mimetic forms. Punnett showed that a selective advantage of 10% would change the ratio from nearly one-half non-mimetic females to less than 1/40th in just 70 generations.[46] A one percent selective advantage would take seven hundred generations to make the change from one-half to 1/40th. Even a selective advantage of only .001% would reduce the population of non-mimetic females to 1/40th in about 1,400,000 generations.[47] Punnett felt that he was surely giving natural selection the benefit of the doubt. Since no change in the ratio between the forms of *P. polytes* occurs in succeeding generations, natural selection must either be so insignificant that its effects cannot be found or it is absent altogether.

The facts of heredity did not coincide with the belief of the selectionists that adaptations are slowly acquired. Punnett and his associates were sure that it was not the case that natural selection was "working, as it were, upon a plastic organism, moulding it little by little to a more and more perfect adaptation to its surroundings."[48] Although that view has a simplicity that is attractive and easy to comprehend, it just was not true, Punnett was sure. The Darwinian view (the "older view," as he called it) did not work out well when pressed hard in particular cases like the study of *Papilio polytes*.[49] It was Punnett's opinion that, since no change was observed in the ratio of recessive (non-mimetic) forms to dominant (mimetic) forms, something must have suspended the dynamic expansion of a newer, better-adapted form and instead permitted the long term equilibrium among the various morphs of *Papilio polytes*. Punnett summarized:

> On this view natural selection is a real factor in connection with mimicry, but its function is to conserve and render preponderant an already existing likeness, not to build up that likeness through the accumulation of small variation so generally assumed. Recent researches in heredity and variation all point to this restriction of the scope of natural selection.[50]

As far as Punnett was concerned, natural selection had no part in the origin of species and little to do with the modification of species toward adaptation. It had only a conserving role in a stable environment, although it might also serve to disseminate adaptations presented to a species by saltation in times of environmental disequilibrium.

The "case of *Papilio polytes*" is particularly fascinating because it was called upon to support yet another kind of evolutionary theory in the 1920's and 1930's--the evolutionary synthesis. *P. polytes* had been singled out as the paradigm example of natural selection's action by Wallace. It had served Punnett to exemplify sudden mutation and, last, it was used by Ronald A. Fisher in 1927 and by E. B. Ford in 1938 to show that polymorphism and mimicry could be developed by natural selection according to the laws of heredity.[51]

Once the phenomena of mimetic polymorphism was announced by Wallace from his findings in the Malay Archipelago, even more remarkable discoveries were yet to be made. A paper by Roland Trimen was read at the Linnean Society of London,

March 5, 1868, announcing a case of mimetic polymorphism which has been repeated as one of the most marvelous ever to be discovered. Trimen's findings in South Africa provided the scientific world with another species on which to center its debates about the evolution of organisms.

Roland Trimen and *Papilio Dardanus*

While "entomologizing in Natal" Trimen came upon some very good examples of Batesian mimicry. He published his findings in the *Linnean Society Transactions* where Bates and Wallace had both published their findings. The issues Trimen discussed in his paper were virtually the same as theirs. The paper, however, illustrates that the focus had begun to shift from the zoogeographical aspects of the phenomena toward the mimetic cases themselves. Although both Bates and Wallace had subordinated the mimetic analogues to taxonomic and zoogeographical issues, Trimen's paper was definitely about the mimics and the models. His paper bore the title "On some remarkable Mimetic Analogies among African Butterflies."[52]

The cases of mimicry Trimen had worked out were so astonishing that it is litle wonder that he treated them as a phenomenon worthy of independent and separate study. The families of butterflies which provided the models for mimicry in South Africa were the same as those which served as models in the New World--the Danaidae and Acraeidae.[53] It had long been known that these groups were part of some strange and apparently anomalous analogies to very distantly related butterfly groups. The French naturalist Jean Alphonse Boisduval had written more than thirty years before:

> C'est une chose bien remarquable que de voir la nature créer à côté les uns des autres l'*Euploea Niavius*, le *Diadema dubia*, et le *Papilio Westermanni*, trois Lépidoptères qui se ressemblent presque complètement par le port, le dessin, et la couleur, quoique appartenant à des genres fort éloignés et de tribus différentes.[54]

Between Boisduval and Trimen, no one had really grappled with the problem of these "recurrent types" in the South African butterfly fauna. Trimen considered that failure an intolerable situation; he wrote: "Entomologists, no less than naturalists generally, appeared content with a child-like wonder at this and kindred factors, and let them pass as things inscrutable."[55] Like Bates and Wallace, Trimen could not seem to let the moment of presenting his new data pass without also giving his personal reflections on the proper attitude toward nature for the student of natural history. The proponents of independent creations of species were obliged to make themselves content to find the ways of nature inscrutable. They must wonder at nature without deeper probing. But, Trimen went on, the view that events and phenomena of nature are all goverened by natural causes and consequences is much more illuminating and allows for questions and permits answers which unite effects with their causes. Trimen wondered why it had been that the notion of cataclysms and convulsions in nature had been expelled from geology but were retained in biology. Sudden creations were still accepted even by many scientific men. He concluded a bit rhapsodically:

> It may, with confidence, be predicted that the day is not far distant when such discordant views of Nature will be unknown in the scientific world; for the progress of discovery proclaims, with ever-increasing force, that the famous axiom, *Natura non facit saltum*, is a truth of universal application, and that it is as impossible to sever the life of the present from that of the remotest past as to impose a barrier between the present and the unknown future.[56]

Trimen had consciously offered his data to the Darwinian side of the controversy. His facts were not neutrally stated; he did not offer them to science for whatever theory the accumulation of facts would eventually bring forth spontaneously.

Trimen had found one of the most amazing and complicated cases of mimetic polymorphism ever discovered when he unravelled the relationships between the forms of the butterfly known to him as *Papilio merope*. It is now usually named *Papilio dardanus*; *merope* is an obsolete synonym. The plethora of names for this species reflected its complicated history. Each of its forms had been found separately and given a species name. Not only were each of the female forms given separate names before it was known which butterfly was the male of the species but P. dardanus (P. merope) was also divided into geographical races. Furthermore, the species testified to its instability by presenting a complex array of intermediate forms. Many clarifications have been made in understanding this species since Trimen's paper, and many nuances have been detected. His success at straightening out the situation is thus all the more remarkable.

His discovery of the connections between the various forms of P. dardanus illustrates the importance of his experiences as both a field naturalist and as a museum taxonomist. Trimen recapitulated the experiences Bates and Wallace had. He had felt the emotion of surprise at removing a Papilio cenea from a net which he had intended for a danaid. At that time he attributed the specimen to an "illegitimate" mating of the "noble Merope and the dusky Danais Echeria."[57]

Almost any collector who found a single specimen he could not account for would think of hybridization as a possible explanation, but two events combined to make him think that hybridization was unlikely. First, Bates had sent Trimen a copy of his paper so that he was conscious of mimetic Heliconiidae in the Amazon. Second, in 1866, Mrs. Barber sent him specimens of Papilio cenea with a request for a male specimen in order to make drawings.[58] Upon searching the collections of the South African Museum as well as his

EXPLANATION OF PLATE IV

Figures:

1. *Papilio dardanus* male, non-mimetic.
2. *Papilio dardanus* female, var. *trophonius*, mimic of *Danais chrysippus*, Fig. 5.
3. *Papilio dardanus* female, var. *hippocoon*, mimic of *Amauris niavius*, Fig. 6.
4. *Papilio dardanus* female, var. *cenea*, mimic of *Amauris echeria*, Fig. 7
8. *Hypolimnas dubius*, var. *mima*, another mimic of *A. echeria*, Fig. 7.
9. *Hypolimnas dubius* var. *wahlbergi*, another mimic of A. *niavius*, Fig. 6.

PLATE IV

MIMETIC POLYMORPHISM IN PAPILIO DARDANUS

(From R. C. Punnett, *Mimicry in Butterflies*, 1915, Plate VIII).

own collections, no males could be found. The combination of the absent female P. merope and the absent male P. cenea was enough to make Trimen sure that he had found a case of female mimicry. His surprise in the field helped him to look for affinity beyond that found in similar coloring. Searches of the full cabinets of the English museums yielded no mates for either P. merope or P. cenea.

Two small entomological papers demonstrate how Trimen gradually attained an understanding of the complex polymorphism and mimicry in Papilio dardanus.[59] In these papers, he stated that he knew that sexual dimorphism may, in some cases, be very great. Many species would be eventually resolved into sexes of a single species. He stated, "Every year's discoveries prove to us that arbitrary entomological decrees of divorce are far more numerous in our collections than we had any idea of."[60] While Trimen was correct that entomologists had been wrong about or ignorant of the relationships between male and female specimens of a single species, the mimetic forms he was studying were another kind of deception. Here there was the appearance of a "divorce" without its reality. The female had taken on the colors of another species. Trimen became convinced of how very good the likenesses can be in mimicry when he found an example of a model catalogued as its mimic in the British Museum.[61] If the staff of the British Museum could mistake the Acraea gea for the Papilio boisduvallianus they might also be very difficult to distinguish in the field. In a little paper, "Aspects of Insect Life in South Eastern Africa," Trimen had described an experience that helped him understand the difficulty one can have in giving specimens their proper identifying tags. Trimen had been observing the male Papilio merope in the company of other butterflies feeding upon lantana in flower.

> Why does the last-named Papilio trouble himself to give chase to one of the many Danais echeria hovering near? See, here they both come back; now's your time to net them together. What a nuisance! You have only netted the Echeria. But stop a minute. Is it Echeria! No! by all that's wonderful, another Papilio, and none other than P. Cyrea of Stoll. When you get home, expand side by side the Danais, the Diadema that you took in the forest, and this Papilio, and then ask any ordinarily intelligent person whether he or she sees any difference between them; he or she will almost certainly reply that you are joking, and that of course they are all the same butterfly. After this, you had better proceed to ask yourself what this wondrous imitation of a common butterfly by two scarce ones of entirely different groups means.[62]

It was apparent that something strange and interesting was to be found by studying these Papilio butterflies. Trimen was of the opinion that, whatever the study would yield in terms of data, it was part of a long history of evolutionary alterations. He utilized the metaphor of the theater--always a favorite with students of mimicry.

> It is gradually becoming acknowledged that the

> accurate description of genera and species, highly important and indeed indispensible as it is, is not the sole end and object of Entomology any more than of other branches of Natural Science. The scientific definition of the peculiar characters of a species is, in fact, only the starting point from which we can safely and surely proceed to investigate its relations, not only to its own near allies and immediate conditions of life, but also to other groups of organic beings, actors like itself in that great drama of Nature which, with not infrequent shifting of scenes and change of performers, had been enacting from the remote dawn of life, and is yet in progress.[63]

In pressing for more emphasis on phylogeny and affinities of living species, Trimen was open to the criticism of those who believed that those matters could only be broached when the species known had been more fully described. Even in the case of the very immediate affinity of the male *P. dardanus* and the female *P. cenea* Trimen was only working from circumstantial evidence. He was basing his claim chiefly on the absence of partners for the forms which looked more like each of them. Trimen did not have as strong a positive case as he might have wished, but he was not one to stay his pen until that remote day, either on matters of theory or on the subject of *Papilio dardanus*.

The eminent entomologist William C. Hewitson raised all the objections--both general and specific--against Trimen's position that could have been anticipated.[64] First, he would not place *Papilio cenea* with *Papilio merope* merely because there was no male like it.[65] Many species were without both sexes in collections: should mates be chosen for them? The fact that P. merope, that is P. dardanus, had a non-mimetic female resembling it in Madagascar should render it improbable that on the African continent the male "indulges in a whole harem of females, differing as widely from it as any other species in the genus."[66] Nor should the fact that a male was seen to pursue a female be seen as prima facie cause to suppose them to be of the same species. Hewitson cited a recent report in the Entomological Magazine of insects from different families being taken together in copulation. He went on in high Victorian style. "It is true that we have of late been introduced to some strange anomalies in the sexes, but to none which bear comparison to this."[67] Hewitson believed that Trimen had merely added insult to imagination with his view that P. cenea was a P. merope female and that the only "rational explanation of these phenomena" was that of Darwin. He quoted Trimen's slur on the "child-like wonder" of entomologists who allow the remarkable analogies to pass as "inscrutible."[68] Hewitson admitted that he was glad to side with the older school. "I must say, and I hope that I may do so without offense to any one, that I prefer the childlike attitude of the former naturalists to the childish guesses of the Darwinian school."[69] Hewitson had used language that would be perfect for caricature should he be proved incorrect later.

The evidence sealing the case of *Papilio merope* was provided by J. P. Mansel Weale and even further evidence, ironically, was published by William Hewitson himself.[70] Weale spent two years working out the life history of *P. merope*. At last someone had reared imagines from larvae found near King William's Town, South Africa. He succeeded in raising the larvae and in observing their pupation and then their emergence as adults. His list of imagines produced from *P. merope* larvae included the male *P. merope*, and the female forms *hippocoon*, *cenea*, and *trophonius*.

With evidence like that obtained from Mansel Weale, Trimen could not be expected to suppress his delight. With obvious satisfaction he wrote:

> It is with reluctance that one contemplates the stretching of Mr. Hewitson's imagination to an "extent of which he is incapable," or the inevitable shock to which his "notions of propriety" will receive, but the evidence now adduced by Mr. Weale is such that the profoundest sceptic cannot explain it away and must allow that the dream had proved to be a true vision.[71]

Dignity does not always characterize discussions of Darwinism and its data! Trimen in particular was prone to flamboyant language and a gloating tone when his opponents were proved in error.

Hewitson must have regretted having spoken so strongly. He wrote a short letter to the editor of the *Entomologist's Monthly Magazine* a few months after the articles by Mansel

Weale and Trimen had appeared in the Entomological Society Transactions. He told there that specimens of Papilio merope and of Papilio hippocoon had been sent to him taken in copulation.[72] Here was another illustration, he acknowledged, that "truth is stranger than fiction." Even with the evidence that P. merope and P. hippocoon were two sexes of a single species, it seemed lacking in verisimilitude that a butterfly with a non-mimetic female form like itself in Madagascar would have several mates in West Africa without even a general resemblance to the male of the species. The data being found in the 1860's was very hard for the naturalist to assimilate who was used to thinking of species as stable, well-defined units. The idea that a species like P. dardanus, then called P. merope, could be made up of a whole cluster of forms was a new and bewildering finding. The question had to be asked, were the P. merope forms on Madagascar the same species as those on the African continent or not? What is the origin of the diverse forms? The Darwinians did their best to dominate the discussion of such questions in the 1860's and 1870's. The absence of mimetic polymorphism in Madagascar confirmed their position that evolution is driven forward by the pressure applied to a species by predators. Predation on Madagascar was quite different from that on the African continent which had an abundant share in insectivorous birds and of protected butterflies to be potential models.

The example of Papilio dardanus became a paradigm example in Darwinian statements of their case. Darwin himself had referred to Trimen's special discovery in the Descent of Man and August Weismann had discussed it in The Evolution Theory (1904).[73] In his book, Weismann had given considerable space to the discussion of the implications of protective coloration research for the evolutionary theory. In a chapter called "True Mimicry," Weismann used many of the same arguments Hewitson had used in his paper for diametrically opposite conclusions.[74]

> There can be no doubt whatever that in these cases a real transformation of the species in colour and marking, and perhaps often, too, in form of wing, has taken place, and that within comparatively recent times--let us say during the distribution of a species which required protection over a large continent, or since the last breaking up of an immune species into local species. Various facts prove this; above all, the circumstance that it is often only the females which exhibit this protective mimicry; and that one and the same species may mimic a different immune species in different areas, but always the one occuring abundantly in that area, and so on.[75]

For Weismann, variation in a species was great enough to allow evolution to proceed in very unexpected and bizarre directions. The special need of female forms for protection had been taken care of in a variable species by mimicry having been developed in the female alone. The destruction of an adult female butterfly meant the loss of perhaps several hundred progeny. Selection had produced, therfore, many changes unique to the female sex. Wallace had held that position for many years before he had Weismann's authoritative backing. Weismann wrote:

> It is, therefore, intelligible that, in species already somewhat rare, the female must first of all be protected, that is to say, that all variations tending in the direction of her protection would give rise to a process of selection resulting in an augmentation of the protective characters.[76]

The adaptive value, then, of mimetic polymorphism was protection. Here was a good case of natural selection in action--at least that was the belief and assertion among Neo-Darwinian entomologists.

During the first decade of the twentieth century, many papers were published on the groups of butterflies containing *Papilio dardanus* and on its various models' groups. Many of the authors of these papers were Darwinians. One important work, published after five years of work, confirmed that mimics and models do indeed fly together as well as inhabit the same geographical areas. This paper, by Guy A. K. Marshall, was published jointly with Edward B. Poulton; it brought together much of the previously published material bearing on the subject of mimicry and evolutionary theory.[77] Marshall and Poulton surveyed the state of knowledge and subjected data to criticism from the persepctive of evolutionary ecology. They worked hard to straighten out the concept of "immunity" which had grown up around the idea of mimicry. They asked, to what degree are distasteful insects "immune" from predation? Bates had believed that unpalatable species were "immune" from the attacks of predaceous insects also.[78] The German entomologist Erich Haase had promulgated that idea in his treatise Untersuchungen über Mimicry.[79] Weismann had adopted the term "immune model" from Haase.[80] The term carried with it the connotation that protected species were beyond the pale of predatory attackers. Poulton and Marshall had set out to learn the degree to which insects' distastefulness to birds granted them protection from other potential predators. Their long study was able to confirm Poulton's contention that parasitic flies and ichneumon wasps feed readily on butterflies which are unpalatable to vertebrates.[81] The subject of mimicry brought forward once again the question of how one would think about animal populations and their histories; even more importantly it raised the issue of how one would interpret natural selection.

Marshall and Poulton's work demonstrated that one should think in terms of statistics and of relative survival rates of specific characters. The dichotomy between "immune" species and non-immune species was an improper distinction. The work of Marshall and Poulton combined field observation and controlled laboratory experiments and gave a new direction and example to subsequent researchers.[82]

Eventually the number of cases of mimicry in South Africa reached the point that a book-length treatment of that one geographical area was warranted. Harry Eltringham published a monograph in 1910 which presents descriptions and color plates of most of the groups of South African mimics and models.[83] Eltringham wrote his book in the spirit of Roland Trimen and with the advice of E. B. Poulton. He believed that at least the South African mimics and models conformed to Darwinian expectations. He too pointed out that the case of *Papilio dardanus* epitomized the South African mimics by manifesting nearly all the phenomena within a single species.

Eltrinham was sensitive to the fact that no bit of his data on *P. dardanus* would escape the hard scrutiny of Darwinian opponents. He wrote of the species:

> The whole life-history of *dardanus* is an illustration of the keen struggle for existence which is met by a development of mimetic modification unparalleled in any other species with which we are acquainted. The most earnest opponent of mimetic theory cannot deny the resemblance of the pupa to a leaf, or that of the male, when at rest to its surroundings. Why are *Amauris echeria*, *albimaculata* and *niavius*, *Danaida chrysippus*, and *Planema poggei* unusually common butterflies, if not by reason of some advantage which they enjoy in their relation to vertebrate enemies, and how is it that out of the unlimited number of patterns which an unstable form like *dardanus* might produce, the principal varieties happen to be those which most resemble these

> slow-flying species? These are problems for which the opponents of mimicry will have to find some much better explanation than those of climatic influence, sexual selection, and parallel variation, before we can see any good reason to abandon Bates' comparatively simple and wholly reasonable explanation.[84]

Eltringham expressed the feeling that by 1910 both evolution and natural selection were on firm ground. The saltationists' case had not been proven; their search for intermediates had not been clearly disconfirming of natural selection, whereas predation by birds had been confirmed. Lamarckian explanations had failed to expalin geographical data which Darwinian explanations dealt with very well. It seemed to Eltringham in 1910 that Batesian mimicry as Bates had described it was both the simplest and best explanatory model for the analogous resemblances called "mimics." But peace was not really at hand because Eltringham declared a winner. The subject of mimicry was not to be settled for a very long time; not the subject of Papilio dardanus and mimetic polymorphism, or mimicry theory in general, or natural selection, or the reality of evolution.

The biology of Papilio dardanus was only well explained in the 1960's, primarily through the efforts of C. A. Clarke and P. M. Sheppard with the assistance of Michael Wells.[85] These researchers have learned that coloration patterns, both mimetic and non-mimetic, are genetically determined and related to the sex of the insect. The presence or absence of the "swallow-tail" is determined by genes independently from the color pattern. The genetic information about mimicry in *P. dardanus* provided by Clarke and Sheppard and summarized by Wolfgang Wickler has been interpreted to mean that mimicry does not appear ready made.[86] It cannot be due to the action of a single gene. The evolution of mimicry is described in terms of natural selection. A change mutation in an unprotected butterfly species that allowed it to escape predation would spread in the population. The degree of similarity between the model and mimic can be improved by mutation and subsequent modification of the gene complex.

Clarke and Sheppard found that mimetic butterflies have a series of modifier genes which regulate the functioning of color pattern genes. They are important since they prevent a confusion of patterns by limiting the number of forms. It is changes in the modifier gene system that allows the deceptive resemblance of the mimic to improve. In crosses between the forms of *P. dardanus*, the effect of modifier genes is not preserved intact. The genes of the modifier system and those of the pattern system are mutually adapted to each other. The combination of systems restricts the variability of forms and provides for gradual heightening of the mimetic effect.

Müllerian Mimicry, 1879

After Bates's descriptions of the analogous resemblances between insects which were palatable to birds and those which were not, his concept of mimicry was extended by Wallace and then Trimen to polymorphic species. By 1870, the concept of mimicry was well known to naturalists. The implications of Batesian mimicry for taxonomy and for geographical distribution work had been repeatedly stressed by the Darwinian naturalists. In 1879, another Darwinian naturalist announced a new kind of mimetic phenomena which he had discovered in butterfly groups, which Bates had also known in the years of his travels in Brazil.

Fritz Müller (1822-1897) was one of Darwin's first German supporters. Müller had published a book entitled Für Darwin in 1864 which drew evidences of evolution by natural selection from detailed studies of marine crustacea, particularly crabs. The abstract of Müller's book stressed his methodology for the English-speaking readers of the Annals and Magazine of Natural History.

> Dr. Fritz Müller has endeavoured...to construct this natural classification or geneological tree for the class of crustacea; and having constructed it, he has deduced from its structure certain necessary consequences. These deductions he has then endeavoured to verify. If they could not be verified, this would be a fatal blow to the Darwinian theory; but if they proved true, they would ufrnish, if not a proof, at least a strong presumption in favour of the theory. Hitherto

> his deductions have been verified; and thus his work presents us with a remarkable example of important results in natural history obtained by a purely deductive method, in opposition to most of the discoveries in that science, which are made by means of a sort of inductive groping.[87]

The methods of deduction and verification which served Müller in his work with crab larvae held good also in his subsequent work with butterflies.[88] He had found that the presumption of evolutionary change by means of natural selection had been a good working hypothesis around which to gather data in Crustacea and Lepidoptera. Of course, Müller did not work with a "purely deductive method" as the writer of the abstract suggested; he was also an expert at "groping" inductively toward interesting findings. The key to Müller's successes as a naturalist was not in the use of a single "pure" method, but in the mixture of ways of thinking and of dealing with data.

Müller was familiar with Bates's work on the protective function of mimicry. Another German naturalist, Oscar Schmidt, had already presented the subject of Batesian mimicry to German readers in Descendenztheorie und Darwinismus in 1873.[89] Müller, however, was in a better position than Schmidt to verify Bates's allegations with his own observations.

Müller persisted in careful observations of butterflies for several years, attempting to ascertain the food preferences of both mimic and of model species. While doing these studies, he found that not only were there resemblances between palatable and unpalatable butterflies, but also resemblances occurred between butterflies in which the "mimics" were very probably as distasteful to most predators as the "models." Müller based his belief that both mimic and model were unpalatable on the fact that they were each from groups which had Batesian mimics patterned after them.

Müller realized that he was dealing with a form of mimetic analogy which was different from the one Bates had described. He had begun to see that the mimetic butterflies which he was studying did not conform to the "laws of mimicry" which Wallace had set. In a little paper in 1878, "Notes on Brazilian Entomology," Müller broke away from prevailing theory in a more important way than immediately meets the eye. In his note, Müller told the readers of the *Proceedings of the Entomological Society* that he harbored a growing distrust of the law of mimicry which asserted that the mimic was always rare but the model was always far more common.[90] His recent experience collecting butterflies had led him to the further doubt of Wallace's second law of mimicry which asserted that the model species was protected in some way not shared by the mimic--either not at all or to a decidedly inferior degree.

During an expedition by canoe on the Itajahy River, Müller had found that his net contained mimics too often for him to believe that they were fewer in numbers than the species which appeared to be their model. He also noticed that these mimics possessed an odor which he found repugnant, leading him to wonder if birds or other predators might also reject the insect as food. The combined force of his waning confidence in Wallace's laws of Batesian mimicry and his field trip to the Itajahy helped Müller to devise a new explanation for these different and difficult cases of mimicry.

In trying to answer the question of how models and mimics are related numerically, Müller had to explain some cases he had found in which the numerical relationship of the model-mimic pair was variable from year to year.[91] The numerical criterion for deciding which species was the mimic and which was the model did not work; by definition the mimic could not become the model and then retreat to being the mimic again. Wallace's other criterion of deciding the mimic from the model on the basis of divergence in color from near relatives did no better in distinguishing the model from the mimic among the insects Müller was studying. Much of the similarity in these butterflies was probably due to their close relationship, that is, to affinity and not only to analogy. Even the criterion of the mimic being relatively more palatable to predators with respect to the model failed as a criterion separating mimic from model.

By every observation, both the _Thyridia_ and _Ituna_ butterflies were protected by unpalatable substances apparently equally. Müller had also noticed that mites and other insects which ravage specimens in collections spare the protected genera (_Euploea_, _Danais_, and certain _Papilio_).[92]

In the novel kind of mimicry he had found, Müller found Batesian explanations of mimicry wanting on two grounds. First, Batesian mimicry did not properly describe the empirical aspects of the mimetic phenomena he was studying. Second, Batesian mimicry theory gave no answer to the question of what the advantage might be to the insect having communal mimetic colors. Since both species of insects were protected from predation by unpalatability, why did several distinct species have the _same_ external appearance?

In trying to answer this question and the related one, "What benefit can one species derive from resembling another, if each is protected by distasteful substances?" Müller turned to his data on the struggle for existence in the insects. He had learned that the major predators on butterflies were birds and lizards, which hunt almost solely by means of visual perceptions. From this observation Müller inferred that common coloration among distasteful insects would be an advantage to each of several species if, and only if, the predators learn by experience which insects make good meals. No benefit would be reaped by mutual resemblance among distasteful insects if the predators have hereditary knowledge of which insects to eat and which to avoid. Any type of hereditary "unconscious intelligence" would negate the advantage of common, conspicuous coloration.[93] Müller wrote: "The experience acquired at the expense of one of them will likewise benefit the other; both species together will only have to contribute the same number of victims which together each of them would have to furnish if they were different."[94] In cases where one species is more rare than the other its share of the benefits of mutual resemblance is increased not only in proportion to its numbers relative to the more common species, but as the square of its number relative to the other. For example, when the numerical relationship between the number of individuals in one species relative to the other is 1:5, the advantage to the rarer species is 25:1.[95]

The number of individuals lost in the process of teaching predators to avoid distasteful prey is constant for any one pattern of coloration. Therefore, there is an advantage to unpalatable species in sharing a recognizable appearance with other species. The uneven distribution of the benefits of mimicry would tend to diminish the more common species and increase the smaller, rarer species. The result would be that the species with a smaller population at the beginning of the effective mimicry would increase its numbers at the expense of the larger species. In the oldest cases of common coloration, the relative numbers of model and mimic species would tend to become equal. Thus, it would be most difficult to state with certainty which species had been the model and which the mimic at the beginning of the mimetic relationship. In addition to *Ituna* and *Thyridia*, Müller named other genera which have species resembling each other so closely that it is impossible to be certain which served as the model for the others.[96]

After Müller's paper on *Ituna* and *Thyridia* had been read and studied by English naturalists, *Nature* published a series of letters discussing Müllerian mimicry.[97] Wallace published a letter stating that he accepted Müller's views as an addition to the Darwinian discussion of mimetic phenomena. Wallace knew that Müller would not get much encouragement in his theories from Bates. Wallace, therefore, seized the opportunity afforded by Müller's recent publication of a new paper to congratulate him on his important contribution to Darwinian entomology.[98] Included in the letter to *Nature*, Wallace gave a calculation of the numbers of butterflies saved from predation by a species involved in Müllerian mimicry.

Wallace had erred in his paraphrase of Müller on the proportions of advantage to each species in Müllerian mimicry. *Nature* was, therefore, obliged to print several letters written to set the record straight.[99] Thomas Blakiston and Thomas Alexander had the last word in the matter.[100] They calculated both absolute and proportional advantages of Müllerian mimics and gave the proper algebraic formulae which both extended and improved upon Müller's original statement that $a^2:b^2$. Blakiston and Alexander stated:

> Let there be two species of insects equally distasteful to young birds, and let it be supposed that the birds would destroy the same number of individuals of each before they were educated to avoid them. Then if these insects are thoroughly mixed and become indistinguishable to the birds, a _proportionate advantage_ accrues to each over its former state of existence. These _proportionate advantages_ are inversely in the duplicate ratio of their respective original numbers compounded with the ratio of the respective percentages that would have survived without the mimicry.[101]

Müller's $a^2:b^2$ only truly described the relative advantage for each of the Müllerian mimics when the number lost because of ineffective imitation was small compared to the total number of individuals in either species a or species b.

Those who attempted to state the degree of advantage quantitatively were only fine-tuning a theory which they considered probably correct. All the attempts to describe Müllerian mimicry in mathematically accurate terms would be to no avail if the theory were discredited by work on other aspects of the insects' biology. If it could be shown by students of animal behavior and psychology that insect predators do not learn which foods are palatable and which unpalatable to them, the entire concept of proportionate advantage to mimetic species would be to no avail, however skillfully stated. This view that insect-eating animals eat few butterflies was stated in the Letters column of *Nature* in 1882 by William Lucas Distant.[102]

Wallace's letter had contained the opinion that an instinct discriminating palatable and unpalatable butterflies was a very unlikely part of the hereditary knowledge of insectivorous animals. Distant wrote to *Nature* to contradict Wallace's opinion based on some studies of domestic poultry which confirmed his opinion that birds have certain instinctive fears.[103] The observation he found relevant was one related to fear in fowls of sting-bearing insects.

William Distant's opinions in these matters had to be taken seriously. He was a widely known entomologist who had special expertise with Lepidoptera. In his book, *Rhopalocera Malayana: A Description of the Butterflies of the Malay Peninsula* (1882-1886), he brought forward additional objections to the mimicry theories of Bates, Wallace and Müller, to supplement his objections published in *Nature*.[104]

The Darwinian who answered Distant's objection to Müllerian mimicry, Raphael Meldola, had been the translator of Müller's papers into English.[105] Meldola, who not only translated, but defended Müller, acted for Müller in a way analogous to the way Huxley served Darwin as his "bulldog."

Meldola, like Huxley, appreciated the Darwinian theory because it used naturalistic explanation exclusively. Müller, who appears never to have considered design as a possible cause for mimetic analogies, held that it was unnecessary to posit any "unknown local causes" to account for mutual resemblances among protected species.

Müller's work on mimicry signals an advance over the work of Bates and Wallace in one very important respect. They both had observed the similarity between related genera. Bates had resigned himself to explaining these similarities as due to "unknown causes"; Wallace had not been able to find a more satisfying explanation. Müller's discovery that insects having similar wing color patterns were protected by unpalatability tied the phenomenon into the Darwinian theory of natural selection and closed another loophole for Lamarckian influences.

In his defense of Müller's theory, Meldola criticised William Distant's use of reasoning by analogy. Distant had used observations of fear, which he asserted to be instinctive, and which were manifested by domestic fowls for bees which can sting, to argue for the instinctive knowledge of wid birds about butterflies being palatable and unpalatable. It was doubted that instinct in domestic and wild birds was the same, and whether knowledge of stings and of unpalatability was in any sense analogous.

Although Meldola's objections were well taken, the Darwinian side of the argument used some dubious analogies of its own. Some of Meldola's arguments seem conspicuously ad hoc. He could do no better than to quote from a letter from Fritz Müller containing the admission that "in the case of birds, I have as yet no direct proof" of hereditary knowledge of palatable foods. Meldola's quotation continued:

> In insects, and especially in bees, my brother Hermann Müller and I have repeatedly observed that they neither know instinctively the flowers which serve to provide them with honey or pollen nor the way in which their booty is best to be obtained. Today, for the first time, a new illustration has been furnished by _Trigona ruficras_ in visiting a _Cypella_ which offers easily accessible honey and pollen, and which the majority of these bees nevertheless could not at first find. Thus, by analogy, the same would occur in birds with respect to eatable insects as in insects with respect to flowers yielding nutriment.[106]

If Distant's analogy between domestic birds' responses to bees and wild birds' responses to nauseating butterflies could be criticised as poor, indeed, the analogy between birds and their prey and insects and flowers was even worse! Müller was focusing on the functional similarity in animal responses to food sources, but the analogy strains credulity without massive support.

The Darwinian defense of its interpretation of the data from Müllerian mimicry was very far from compelling.

Both Müller and Meldola were working with the data of mimetic insects from inside the Darwinian "world-view." Their method of arguing is markedly deductive and, therefore, unconvincing to anti-Darwinians. It was also marred by unsound or ambiguous use of analogous reasoning and by evidences of insecurity in dealing with data presented by partisans of anti-Darwinian views. Distant and Meldola were really talking past each other rather than talking to each other.

Museum-Made Mimicry

The years following Müller's paper on mimicry between protected, unpalatable butterflies were marked with heated debate about the meaning of the mimetic phenomena. The Darwinians, who conducted much of the research on mimicry, were eager to see their branch of entomology recognized as having scientific stature and merit. Different naturalists employed various means to this end. One such naturalist, Raphael Meldola, tried to relate the predatory behavior of birds, which was important for mimicry theory, especially in Müllerian mimicry, to Haeckel's biogenetic law, which had commanded much attention. Meldola attempted to show that Haeckel's doctrine of recapitulation had implications for instinctive behaviors, but only for behaviors which were of vital importance to animals.[107] Meldola was trying to make the case against William Distant that instinctive fear of bees could become hereditary and expressed by very young birds, but knowledge of palatable foods would not. Meldola sought the sanction of Haeckel's "Law" for behaviors relating to mimicry:

> If instinct is habit acquired during the former experience of the race, and accumulated and fixed by heredity, we may fairly expect that an animal, in the course of it mental development (psychological ontogeny), would pass through the stages of inexperience which were gone through by its ancestors in the course of their evolution. Only where an "instinct" was of vital importance to the young would it become fixed upon the early stages of growth by the law of inheritance at corresponding periods. The instinctive fear of bees shown by Mr. Spalding's turkey may be a case in point. But, if, as I believe, the knowledge of a nasty butterfly is not a matter of vital importance, there is no reason why young birds should know such species antecedent to experience.[108]

Meldola's opponent, William Distant, was unconvinced by Meldola's alliance of Haeckel's biogenetic law and unimpressed with his logic.[109] Distant referred to Meldola's paper as belonging to the armory of "forensic biology." He called for a return to careful scientific work, as distinguished from the multiplication of "mimicking theories" of a "purely speculative character."[110] Distant had some reasons to remain unconvinced that Müllerian mimicry was not, after all, only the result of confusions in taxonomy rather than common warning colors in protected genera. The addition of new theories of mimicry and attempts at analogizing the phenomena of organismic zoology to the law of recapitulation, which was worked out for embryology, did not help the Darwinian position very much. If anything, it made the Darwinians seem more fanciful and less willing to do the real work at hand in straightening out confusions in the classification of difficult genera like Müller's Euploea.

Some Darwinians turned to a more profitable method of confirming mimicry theory. John Jenner Weir began a series of experiments testing the responses of birds to different foods. He kept notes on which insects were accepted and which rejected.[111] The experiments had been started at the suggestion of Alfred Russel Wallace as tests for Batesian mimicry in 1868. His results confirmed Bates's theory, since he found that conspicuous larvae, known to him to be from protected, unpalatable species, were rarely even examined by the birds. Bates's theory was upheld by the birds' behavior, since the abstinence from eating unpalatable insects indicated the birds' responsiveness to warning coloration. Müller's view of mimicry was not confirmed by Jenner Weir's experiments.

Jenner Weir had concluded that the birds had some kind of instinctive knowledge which gave them information as to the larvae which were edible and which were inedible. Although Jenner Weir was a convinced Darwinian, he was not able to confirm Müller's theory that mutual resemblance of protected genera had evolved through the learning experiences of naive young birds.[112] It seemed to Jenner Weir that Müllerian mimicry was struck down on its presumption that birds must learn their appropriate food sources by trial and error without the aid of hereditary "unconscious cerebration."[113]

Jenner Weir, however, was not the only Darwinian entomologist to have serious doubts about Müller's additions to mimicry theory. Even Henry Walter Bates had questions about whether or not Müller had identified a new kind of mimicry or only raised another taxonomic puzzle in a very difficult group.[114] Bates had been present at the meeting of the Entomological Society at which Meldola's translation of Müller's paper, "*Ituna* and *Thyridia*," was read. The minutes of the meeting record Bates's reaction. The *Proceedings* contains this statement:

> He [Bates] could not see that Dr. Müller's explanations and calculations cleared up all the difficulties. The numerous cases where species are themselves apparently protected by their offensive secretions evidently mimic other species similarly protected still form a great stumbling block.[115]

Bates remembered the butterflies Müller described from the years of his travels and summarized the phenomena found in the Heliconiidae for the members of the Entomological Society. Bates accented zoogeographical issues which Müller had not studied in the same detail Bates had. He noted that the species change every few miles along the Amazon, "as if by the touch of the enchanter's wand."[116] Despite the changes of species, there was still a remarkable constancy of color and form found among species from different genera. Acknowledging that his own zoogeographical work had given him the needed clues to understand mimicry, Bates stated that he knew that he had not solved all the questions regarding the Heliconiidae. It did not seem to him that Müller had solved them either. The problems of classification were extremely great; Bates stated that he had seen some half-dozen species of different genera classed and named in collections as one species. Müller's paper seems to have merely confirmed Bates's own opinion that the subject was excessively complex and would yield only slowly to the combined methods of taxonomy, zoogeography, and evolutionary theory.

Bates's uncertainty and lack of enthusiasm for Müller's theory of mimicry was of considerable interest to entomologists and other "Darwinian-watchers" of the late nineteenth century. His reaction was particularly important since he was the first Darwinian theorist on mimetic phenomena, besides being an expert on the group both as a field biologist in Brazil and as a taxonomist in his later years. Any rifts in solidarity of opinion among natural selection advocates were certain to be noted by proponents of other views, even when it was only the mild dissension of withholding one's endorsement.

Bates's cool reception of Müllerian mimicry was pointed

out by Alpheus Spring Packard, one of the most prominent of American Neo-Lamarckians.[117] Packard remembered that Bates had allowed for the possible action of agents of change in addition to natural selection. In the creation of local varieties, Bates had considered the possibility that climatic and other physical conditions had altered the species through the inheritance of acquired characteristics. From the Neo-Lamarckian perspective, Bates was "less narrow" and more "broad and sound" in his discussions of mimicry than later writers, like Müller, who explained the origin of mimetic phenomena on the basis of evolution by natural selection alone.[118] Müller had refused to admit that mimetic resemblances between different species might be due to evolutionary convergence resulting from the action of physical conditions on them. It was not environmental influence that made different species alike, but the agents of natural selection--the bird predators of insects. Packard believed that Müller and others would find the bird predators elusive. He believed that birds rarely ate butterflies at all.[119] Contrary to the assertions of the Darwinians, the alleged agents of selection were not doing much selecting if they were doing any at all. Packard challenged the Darwinians to show that birds eat butterflies. It was a more difficult field problem than it might seem, because the Darwinians must show not only that birds take butterflies as prey, but that they do so with enough relish to make changes in the species. Packard's challenge was added to Distant's, to show that birds learn which insects are good to eat, and demonstrates that the onus of proof was still with the Darwinians. That was particularly the case with the problems of mimicry involving the behavior of wild birds.

The Darwinian position on mimicry, both Batesian and Müllerian, was doubted on many grounds during the 1890's and the first decade of the twentieth century. A series of complaints was lodged against mimicry theory during that period. One example of the rising doubt about mimicry is found in a letter to the editor of Nature in 1898 from George F. Hampson, a worker in the Entomology Department at the British Museum.[120] Under the heading "Protective and Pseudo-Mimicry," Hampson complained about the "rapidly swelling ranks of museum-made mimicry." Scholars had become unscholarly in their haste to adduce fresh cases of mimicry.

> Protective mimicry entirely depends on community of habit and it used to be considered that accurate field observations were necessary to establish it. Now that it has been degraded to the matching of specimens in a drawer like ribbons in a store, being a nice easy subject to philosophise on and entailing a minimum of work, it has lost all claim to serious consideration.[121]

Of course, Hampson was correct that superficial similarity in color patterns must not be confused with mimicry without corroborating field investigation. These field researches were both difficult and immensely time-consuming.

Once again it was the Darwinian zoologists who undertook the laborious field studies. The status of research on mimicry is summarized for the year 1908 in the Entomologist's Record in a section on the year's publications and discussions by members of the South London Entomological and Natural History Society.[122] The members of the society had been greatly interested in a paper by Guy A. K. Marshall entitled, "On Diaposematism with Reference to some Limitations of the Müllerian Hypothesis of Mimicry."[123] The report notes with gratitude the care with which Marshall had observed insect habits. At last someone had given good information that mimics fly with their models on the same ground and at the same time. Empirical field work promised to revitalize interest and restore credibility to a field of research which "has made most lepidopterists a bit tired of this museum-made mimicry."[124] The report continued:

> What a lot of time, trouble, and paper would have been saved had Mr. Marshall written us the facts of the life-habits before the supposed mimicry of the species had been dealt with, rather than have had the suppositious side of their mimicry written up first, and the facts of their life-habits so much later! It is advisable to keep well in mind the critical, although offhand, statement made by one of our fellows after a lengthy discourse on the "seasonal dimorphism" of some Central American Callidryads--"Ah, yes, if they are the examples in the 'X' Museum, I think I took them congregated round the same puddles on the same day." One wondered then, and still wonders,

how far merely museum theories are ever right.[125] In the minds of some zoologists, mimicry theory had lost its close connection to observations and to confirmable "fact." George Hampson's letter in 1898 and the report from the South London Entomological and Natrual History Society for 1908 agreed that the reputation of mimicry studies was in steep decline. It had become associated in the minds of many naturalists with casual matching of specimens in museum collections.

The critics of mimicry theory were making some valid points, and orthodox Darwinians were hard-pressed to satisfy them. Men like E. B. Poulton, Guy A. K. Marshall, and F. A. Dixey were doing their best to give some compelling answers to the objections to both Batesian and Müllerian mimicry. The work of the Darwinians in this period combined field and laboratory research and helped mimicry studies to improved reputation among entomologists. The Darwinians realized fully that it mattered very much whether or not the agents of natural selection (birds) were selecting and they provided major studies to demonstrate that they were.

There were a few areas of disharmony within Darwinian circles which complicated the discussions of mimicry and prevented the development of a consensus of Darwinian views. Students of Müllerian mimicry began a reassessment of data which had been used to illustrate Batesian mimicry. The publications by Frederick A. Dixey from 1894 to 1897 reinterpreted much of the data supporting Batesian mimicry as actually supporting Müllerian mimicry.[126] Edward B. Poulton attempted, with limited success, to emphasize that the two views were not competitors or mutually exclusive. In Essays on Evolution, 1908, Poulton acknowledged that Müllerian mimicry had recently become the predominant view of mimicry, that some examples had been moved from the Batesian to the Müllerian category as more information about the butterflies and their predators had become known.[127] The progress of Müllerian mimicry toward acceptance by naturalists had been slow, but it surpassed Batesian mimicry in the early twentieth century. In his Huxley Lecture for 1905, Poulton ascribed the slow pace at which naturalists had accepted Müllerian mimicry to the fact that the Batesian explanation of mimicry had been around for seventeen years and had considerable hold on the minds of those who cared about such matters. Poulton pointed out, moreover, that Müller's paper "Ituna and Thyridia" was based on rather plain butterflies and was presented without color illustrations.[128] Poulton's point deserves emphasis; the visual component is extremely important in demonstrating mimetic resemblances.

The increasing rivalry between the Batesian and Müllerian theories of mimicry after 1890 tended to weaken the solidarity of the Darwinian position. Every problem within

Darwinian ranks was met with several attempts by anti-Darwinians to dispute mimicry theory as the Darwinians understood it. The connections between mimicry and natural selection were clear to everyone. Therefore, persons who had aversions to the theory of evolution by natural selection were obligated to dispose of mimicry theory which seemed to support natural selection.

Anti-Darwinian Objections and Refutations

Among the major critics of mimicry theory, there is one who exceeds the rest in his persevering, dogged attempts to refute Darwinian theories. He had written a critical review of the *Origin of Species* in 1860; he had opposed Wallace's view of the relationship between bird nests and bird coloration. In 1870 he was prepared to try to refute Darwinian claims to explain mimetic coloration in insects.[129] Neither the major nor the minor premises of evolutionary theory were acceptable to Andrew Murray and he had had practice in debates with Darwinians.

Murray tried to show that the phenomena alleged to be due to mimicry might have been caused by hybridization between species.[130] He believed, furthermore, that natural selection was an incompatible explanation for the very data to which it had been applied by Darwinian mimicry theorists. He stated at the outset that opponents of Bates's and Wallace's view of mimicry made a mistake by arguing with Darwinians on their own terms. Murray wrote:

> When I first approached the subject I did the same; but the longer I live and the more extended my experience becomes, the more surely do I find that when a theory looks shaky and unsound, the place to look for the flaw is not in the upper story, but in the basement. It is in the foundation that the crack will almost invariably be found. I am sure it is so here.[131]

In his view the crack in the Darwinian basement was natural selection. He was of the opinion that the isolation of hybrids between species could give rise to new species. Thus, evolution occurred by the isolation and by hybridization rather than by natural selection.[132] Murray badly misrepresented Batesian mimicry in his paper by claiming that Bates believed the Heliconiidae were protected by a bad smell. He correctly stated that insectivorous animals hunt by sight and not by smell, but he was badly in error to believe that Bates asserted a contrary opinion. Bates had clearly stated that any bad smell an insect might have was a clue to its having the protection afforded by an unpalatable taste. The conspicuous coloration associated with unpalatability served the insect as a signal. Murray went on in a manner that was most unflattering to Bates:

> My object is not so much to show that a friend and entomological brother has been seduced by a "bad smell" to go on a wrong scent, like a good dog after a red herring, but to find out the true explanation of the phenomenon.[132]

Although many of the Darwinians must have privately objected to Murray's _ad hominem_ remarks, especially to the "canine" metaphor, none of them mentioned it in their letters to _Nature_.[133] Murray's objections to mimicry theory are chiefly interesting as a sample of a concentrated barrage against natural selection as it is understood in connection with the adaptive coloring of animals.

Much more serious objections were lodged against the Darwinian view of the evolution of mimetic resemblances by the selection of individual variations. These raw materials of evolution seemed to some to be too small and the process so gradual as to become incredible. Many steps toward mimicry would have to be produced by chance variation before the resemblance to a protected model would become useful for deception.[134] Even those who might grant that natural selection might be the agent of perfecting a reasonably good mimetic resemblance could not also grant that it could have been produced from its beginning by natural selection alone. A representative of this objection to the evolution of mimicry by natural selection, Alfred W. Bennett, said that a minimum of one thousand steps would be required for the _Leptalis_ to pass for an _Ithomia_.[135] He estimated that at least twenty of those steps had to be accumulated by chance before any subsequent steps could be taken by natural selection of useful variation. The chances of those first twenty steps occurring he calculated at one in about ten billion. Like many other people of his time, Bennett believed that some "Law" must be in operation to govern the production of variations and to render change directional in its orientation. It was the only solution he could find to the problem of the inherent improbability of so many steps toward an end occurring by chance. Modern biologists have found since Bennett's time that populations are much more rich in variation than they were thought to be in the nineteenth century. Bennett's conclusion was that the naturalist must admit that an "Unconscious Organizing Intelligence" plays a role in producing directional variation.[136] He stated:

> And if this inherent innate power of change is admitted, it at once harmonises the tendency to variation which exists in all created beings, with the perpetuation of those forms best adapted to resist the struggle of life, and lends to natural selection the assistance of a fellow worker far more powerful and of more universal operation.[137]

Bennett rejected the view of evolution associated with Alfred Russel Wallace's work, which seemed to him to require an absolute denial of design in nature. This extreme form of Darwinism, he said, was "the carrying into Natural Science of the Hobbesian principle of self-love."[138] The Darwinian position allowed no room for any but selfish motivations; it negated the lesson that "All are but parts of one stupendous whole/ Whose body Nature is, and God, the Soul."[139] The essay by Bennett was written as an answer to Wallace's work on mimicry in the Contributions to a Theory of Natural Selection, where Wallace had extended the function of natural selection beyond what Bennett thought was the reasonable province of its operation.[140] Since mimicry had been chosen to exemplify the action of natural selection, it was there he would offer his rebuttal.

Bennett's mathematical calculations were disputed by Darwinians who felt that not nearly so many as one thousand steps would be required to take an insect from a non-mimetic to a mimetic condition. Although there were many questions about mimicry raised by people interested in variation like Bennett or in hybridization like Murray, these were not the most serious challenges to the Darwinian theory of mimicry and natural selection. One of the most important papers casting doubt on the Darwinian interpretation of mimicry was written by Alpheus Packard, who presented evidence from observations on animal behavior.[141] He had extracted from the published literature and from personal communications dozens of comments by ornithologists, entomologists, collectors and observers of every kind indicating that birds eat few, if any, butterflies. Packard compiled the testimonials of well-known scientists like Samuel Scudder of Harvard, who said that he had never seen a butterfly in a bird's beak. His statement was repeated by William Morton Wheeler.[142]

Packard pointed out that Bates never stated that he had observed birds in pursuit of butterflies, or eating them.[143]

Packard's paper was a threat to Darwinian interpretation of mimicry because he made it clear that the circumstantial evidence of the existence of mimicry in butterfly species does not warrant the unsubstantiated belief that birds are the agents of natural selection.[144] The fact that people who watch birds do not see them in pursuit of butterflies contradicts the Darwinian hypotheses of mimicry. He had, it must be admitted, a very large dossier of statements against the view that birds eat butterflies as a normal part of their diet. As a discipline, animal behavior was rarely the subject of much systematic study. What was not known about animal behavior was hurting the Darwinian position.

Bates and Wallace had shown that natural selection could render the color patterns of one butterfly indistinguishable from that of another species. It was the aspect of mimicry other than the evolution of coloration patterns that gave them the most difficulty in answering challenges. Observations of enough birds in the wild to convince Packard that birds do eat butterflies were not immediately avaliable. Some of the best work had been done to corroborate the Darwinian position by E. B. Poulton, who had shown that birds and lizards would eat imago forms of butterflies and that they have definite taste preferences which were inversely correlated with conspicuous coloration.[145] But Jenner Weir, A. G. Butler, E. B. Poulton and their associates failed to show to Packard's satisfaction that the results with captive animals would be the same as with wild animals. Although caged animals might give some indication of what a wild animal might do, it was not proof.

Alpheus Packard doubted every aspect of Darwinian mimicry theory. Not being a naturalist to quail at the logical conclusion of his doubts, he tasted a Monarch butterfly which is the model of the mimic Viceroy and he stated that they were not particularly unpalatable! He wondered on the basis of his own taste-tests if the unpalatable models might not be so unpalatable after all, as Bates, Wallace, Müller, and others had attested.[146]

Although most of his remarks were critical of Darwinian theories of mimicry, Packard had some positive assertions to make connecting the phenomena of mimicry to a Neo-Lamarckian view of evolution. The similarity of coloration observed in model and mimic species was a result of their living in a common environment. Packard accepted the work of Abbott Henderson Thayer, who had interpreted the color patterns called warning colors by "ultra-Darwinians" (i.e., Wallace, Weismann, and Poulton) as actually concealing colors.[147] Quoting indirectly from Thayer, Packard said, "The prevalent idea of protective mimicry makes an animal appear to be some other thing, whereas it makes it cease to appear to exist at all."[148] Packard agreed with Thayer that any colored pattern is less conspicuous than a bright monochrome.[149] For a Neo-Lamarckian, it is easier to explain general blending coloration on the basis of the influence of sunlight or other physical factors than to explain genuine mimicry. Insofar as possible Packard explained cases of mimicry as cases of general cryptic coloration.

During the first fifty years of Darwinian controversy, 1859-1909, mimicry theory made many advances as observational data expanded. It did not do so unopposed. Until there was some clearer idea of the nature of variation, its usual size, and its relationship to heredity, the saltationists could press their claims that mimicry was a case of the parallel mutations having appeared in distantly related animals. Neo-Lamarckian views attracted many adherents with the explanation that mimetic similarity results from physical similarity in the conditions under which animals live. Both the Mendelians and the Neo-Lamarckians, of course, accepted evolutionary change, but there were others who doubted the entire theory, questioning evolution as well as natural selection. Darwinism was a minority view throughout the period studied here, especially among "non-naturalists"--particularly experimental geneticists. Since mimicry theory had been heralded as offering proof for natural selection, it was singled out for severe attack.

Conclusion

During the decade following Bates's publication describing mimicry in neo-tropical butterflies, similar phenomena were found in the eastern as well as in the western hemisphere. Each new discovery was accompanied by a Darwinian explanation relating the analogous resemblances of palatable butterflies to butterflies not eaten by birds to "Mr. Darwin's principle" of natural selection. During the first ten years of mimicry theory, Wallace found and described mimetic polymorphism and established "laws" or criteria for mimicry, providing a framework for understanding the relationships of mimic species to model as well as to insect predators. Roland Trimen's discovery in South Africa of a butterfly species with both mimetic and non-mimetic female forms added to the general amazement at a phenomenon which was already striking both visually and intellectually. The Darwinian workers had every reason to be optimistic about the usefulness of mimicry as a phenomenon supporting natural selection. Despite the protestations of William Hewitson about the violations of the Victorian sense of propriety, Darwinian mimicry theory was being progressively supported by new findings.

Some of the most serious troubles for the Darwinian students of mimicry were rather late in arising. For some years, Wallace's interpretation of mimetic polymorphism in

Papilio polytes was not severely challenged as an exemplar of natural selection. Some forty-eight years elapsed between Wallace's paper on the Malayan Papilionidae and J. C. F. Fryer's use of the species as an example of discontinuous variation. Another pair of species which had been described as model and mimic evolved by natural selection was reinterpreted by opponents of Darwinism. The North American Monarch and Viceroy butterflies had been described by Benjamin Walsh and Charles Valentine Riley in terms of their evolution through natural selection. Alpheus Packard interpreted the similarities in the butterflies to mean that both species were responding to similarities in their physical environment. These reevaluations of work done by Darwinian naturalists was part of the eclipse of the Darwinian theory of evolution by natural selection during the last years of the nineteenth century and the first years of the twentieth century.

After Müller's paper on a new kind of mimicry in 1879, the discussions of mimicry were not so clear and certain as earlier papers had been. In spite of the fact that both Batesian mimicry and Müllerian mimicry were equally explicable in terms of the struggle for life, adaptation, survival and natural selection, there was nothing approaching consensus of opinion among Darwinians about their relationship. Müller had described a type of mimetic resemblance for which Wallace's laws did not work. Besides this, Bates himself could never quite accept Müller's view of resemblances evolving among unpalatable genera. In the 1890's the work of Frederick A. Dixey seemed to argue for the expansion of Müllerian mimicry to the exclusion of almost all examples of Batesian mimicry in butterflies.[151] Although the proliferation of separate Darwinian explanations of diverse mimetic phenomena entailed some uncertainty among the advocates of Darwinism, the most severe challenges to mimicry theory came from outside the circle of naturalists.

The geneticists led the arguments against mimicry theory and natural selection. Very few Mendelians between 1900 and 1930 believed that natural selection was an important factor in evolution. As a rule, geneticists felt that natural selection was too slow to be the exclusive mechanism of evolution and its steps too small to account for directional changes of species. As late as 1937, A. Franklin Shull wrote a critique of Darwinian views of mimicry in Science:

> There is no common ground for discussion until the fundamental question, what is involved in judging mimicry, is settled. Examples mean nothing unless it is understood what they show. A critical biologist will hold that these judgments lie mainly in two fields, genetics a and animal behavior--the former relating to the origin of mimicry, the latter to its current value to other mimics.[152]

It was really only subsequent to the efforts of such workers as Ronald Fisher, E. B. Ford, and G. D. Hale Carpenter in the late 1920's and 1930's that Darwinian principles began to be widely acknowledged to adequately explain both the origin and continuation of mimetic resemblances in butterflies.[153]

Critics of mimicry were also found among students of animal behavior well into the twentieth century. For a time it was asserted that butterflies were not eaten by birds and then the opposite was claimed, that birds ate butterflies absolutely indiscriminately. W. L. McAtee was the primary proponent of the view that conspicuous coloration was not a signal of unpalatability since birds found all insects edible.[154] Therefore, mimicry could not have the effect of shielding insects from predation. Eventually important flaws in McAtee's experiments were pointed out by Darwinians, but animal behavior studies on mimicry were fraught with difficulties both in methodology and in philosophy.

Field observations were not very systematic, despite the good efforts of naturalists like Frank Finn in India.[155] For a long time there was no mass of inductive facts from which to work; so the Darwinians were forced to rely in part upon deductions from their theory. Scientists who were not committed to the same theoretical framework were won over to the Darwinian point of view very slowly.

The subject of mimicry made its way even into the popular press on a few occasions, especially in connection with a display of models and mimics at the British Museum of Natural History in 1906. A reporter from Punch captured the confusion typical of much of the writing about mimicry in an article called "Reciprocal Advantage" under the general heading, "Moral Reflections at the British Museum."[156] The article's title, "Reciprocal Advantage," indicates that the reporter had seen a display of Müllerian mimics; the writer nonetheless described the phenomenon of Batesian mimicry. Even the author's errors are indicative of the state of understanding. He wrote:

> So far as I can follow it, the situation is something like this: There is a Nice Butterfly (whom we will call A.) who is uncomfortably conscious of being regarded as a delicious tit-bit by every bird or lizard with any pretensions to palate.
>
> There is also a Nasty Butterfly (hereinafter referred to as B.) whom even the least particular lizard or bird will, after a single experience of his peculiar flavour, take uncommonly good care never to touch again.
>
> So says A. to himself (or else Nature says it for him. I am too unscientific to know which): "If I could only make myself <u>look</u> as nasty as B. <u>is</u>, all the birds and lizards would let <u>me</u> alone!" Which, by patience and perseverence, A. gradually contrives to do. Well, I will admit that this idea of A.'s is not without a showy ingenuity, even if it is wanting in true reciprocity and consideration for B.'s interests. What I fail to see is that, even from A.'s point of view, it is really such a masterly stratagem.
>
> The calculation evidently assumes that the inexperienced lizard (or bird) would come across B. <u>first</u>. But why is this so certain? Why should not A. be the first victim? Then obviously, such bird (or lizard) would infer that both A. and B. were equally good eating. I may be told that any such impression would infallibly be corrected as soon as the lizard (or bird) tried to tackle B.--that it would then wipe its mouth (or bill) with surprised disgust, and make

> a mental note to avoid both B.'s and A.'s in the future. Again, _Why_? Surely it is quite as likely that the bird (or lizard) would merely look upon the _contretemps_ as a regrettable incident, a rare exception to the general rule. After all, a person who has had the ill-luck to absorb a dubious Anglo-Dutch does not necessarily forswear all natives from that moment. He may pass through an interim of suspicion, but eventually, in ninety-nine cases out of a hundred, hope will, as it generally does, triumph over Experience. His trust in oysters will return to him.
>
> No, Scientific Experts may extol the wisdom of this so-called Reciprocal Arrangement if they choose:--for my own part, I have the gravest doubts whether it is found to work out well in practice. Depend upon it, when any insect sets out with the idea of being too clever by half, it is tolerably certain to be itself the first it succeeds in taking in! That, in all events, is the Moral _I_ drew from the case in point.[157]

No clear distinction between natural selection and the inheritance of acquired characteristics was made in _Punch_. Alvar Ellegard has stated that the magazine literature of the time very often associated the views of Darwin and Lamarck.[158] The author of the lampoon hit upon a number of rather ambiguous points in both mimicry theory and in evolutionary theory in 1906. The writer was correct that it is very important to know whether hope triumphs over experience in birds (or lizards) as it does often in people. He was also right that a great deal depends on whether or not the lizard (or bird) is able to regain its "trust in oysters." Some of these questions were still open.

Without having seen the exhibit it is not possible to

be certain whether the viewer missed the point, deliberately misconstrued the display for humorous effect, or if the display did not succeed in making the point that advantage in mimicry is only conferred by a statistically better chance of survival. Statistics involved new and unfamiliar ways of thinking in relative rather than in absolute terms. The controversy about Müllerian mimicry and its advantage had involved complicated calculation which was not very common at that time in natural history literature. Many biologists distrusted mathematics as artificial and abstract; figures and calculations could be used to prove opposite points. Alfred Bennett, for example, had shown by using mathematics why mimicry was impossible, and Wallace, Meldola, Blakiston and Alexander had demonstrated its likelihood in similarly mathematical terms. William L. Distant reflected common doubts as to the relevance of calculating "reciprocal advantages" in his criticism of Meldola's numerical arguments:

> If figures could prove biological hypotheses [Meldola's calculations] would leave nothing to be desired. However, 'Nature' does not readily unfold herself to this method; and it must not be forgotten that Kramer has used the same artificial means in an anti-Darwinian sense.[159]

It was a long time before statistical, relativistic thinking became easy, or even possible, for naturalists who had been trained before the 1860's or 1870's.

The extension of Batesian mimicry to many examples

gathered from all over the world did not carry the day for the Darwinians in any complete sense though they hoped it would. It did serve them well as a means to teach other people how natural selection can change species. The data which best supported the Darwinian theory of evolution by natural selection was too easily assimilated by both the saltationists and the Neo-Lamarckians.

It is ironic that at the end of the period emphasized here (until about 1909), the best examples of mimicry as explained only by natural selection were those examples made famous in the textbook of the natural theologians, William Kirby and William Spence.[160] Kirby and Spence described mimicry of bumble-bees by species of flies. Of course, for them the resemblance of a fly to a bee was an indication of benevolent design. The Darwinians had reinterpreted the resemblance between members of different orders as a case of Batesian mimicry whereby the unprotected fly evolved external likeness to an insect which is recognized as inedible by birds. The Mendelians, though willing to admit that nature takes leaps, were hard-pressed to explain how single mutations could explain extremely good resemblances between such remotely related insects as represented by two distinct orders. Similarly, the Neo-Lamarckians found it difficult to explain the mimicry of flies to bees or of moths to bees. The different orders had divergent life-styles and were influenced therefore by both different physical and biological conditions. Although the Lamarckians were easily able to explain analogous resemblances in species which were related taxonomically and functionally, they were not able to answer the Darwinians so quickly when the species represented different orders.

Mimicry between organisms far removed from each other, like that between the bumble-bee and the robber-fly, was explained first by the natural theologians and then reinterpreted by Darwinians. By the end of the nineteenth century very little was being written from the point of view of natural theology. The period ended with the Darwinians in best control of data provided to them by natural theologians. Much of the new data found after the publication of the _Origin of Species_ was shared by those other evolutionary sects which were live options in biology at the end of the nineteenth century and the beginning of the present century. Mimicry was to become once again a paradigm example of evolution by natural selection only after the synthesis of Darwinian theory of evolution by natural selection with the new science of genetics.

Notes--Chapter III

MIMICRY WORLD-WIDE

1. Henry Walter Bates, _The Naturalist on the River Amazons_ (London: John Murray, 1863) p. 18. Many of the details of scientific interest were cut out of the second edition of the Naturalist when it was reduced to a single volume. It retained the same scope, but with a smaller number of specific examples.

2. Bates, "Contributions to an Insect Fauna of the Amazon Valley. _Lepidoptera: Heliconidae_," _Linnean Society Transactions_ 23:495-566, 1862.

3. Bates, _Naturalist_, 1863, 1st ed., pp. 208, 260-261, 263-265.

4. Bates, _Naturalist_, 1863, 1st ed., pp. 19, 21-22.

5. Morse Peckham, editor, _The Origin of Species by Charles Darwin. A Variorum Text_ (Philadelphia: University of Pennsylvania Press, 1959), pp. 666-668. 145.1:d and 145.22:d. Darwin relaxed his superlatives in his sixth edition so that the sentence stated, "I allude to the wonderful manner in which certain butterflies imitate, as first described by Mr. Bates, other and quite distinct species." 145.1:f. (Hereinafter referred to as _Variorum_.)

6. Peckham, _Variorum_, p. 669. 145.23:f.

7. Benjamin D. Walsh and Charles V. Riley, "Imitative Butterflies," _American Entomologist_ 1:189-193, 1869.

8. Thomas Belt, _The Naturalist in Nicaragua_ (London: John Murray, 1874), pp. 6-8, 108-109, 314-321, 382-385.

9. Alfred Russel Wallace, "On the Phenomena of Variation and Geographical Distribution as Illustrated by the Palilionidae of the Malayan Region," _Transactions of the Linnean Society_ 25:1-72, 1865-66. (Read March 17, 1864.) Also, Roland Trimen, "On some Remarkable Mimetic Analogies among African Butterflies," _Transactions of the Linnean Society (Zoology)_ 26:497-522, 1870. (Read 1868.) (Wallace's paper hereinafter referred to as "Malayan Papilionidae.")

10. Fritz Müller, _Für Darwin_ (Leipzig, 1864). Muller's work was known in England as _Facts and Arguments_

for *Darwin*. Translated from the German by W. S. Dallas (London: John Murray, 1869).

11. Fritz Müller, "*Ituna* and *Thyridia*: A Remarkable Case of Mimicry in Butterflies," translated by R. Meldola, *Proceedings of The Entomological Society of London for 1879*, pp. xx-xxix.

12. Wallace, "Malayan Papilionidae," 1865.

13. Wallace, "Malayan Papilionidae," 1865, p. 5.

14. Wallace, "Malayan Papilionidae," 1865, p. 14.

15. Wallace, "Malayan Papilionidae," 1865, pp. 5-10, and 19-23.

16. Wallace changed the nomenclature of the species considerably. *Papilio polytes* is the name for the species used in modern studies of the group. A synonym for it in the nineteenth century is *Papilio pammon*. Wallace classified as two female forms of *Papilio polytes* butterflies which had been given separate species names. The old names were retained as designating the female forms. He used the names *romulus* and *polytes*.

17. Wallace, "Malayan Papilionidae," p. 7.

18. Wallace, "Malayan Papilionidae," pp. 9-10.

19. Wallace believed that the absence of intermediates might be due to their having been eliminated by natural selection during the process of evolutionary divergence. The geneticists of the early twentieth century understood the absence of intermediate forms as an indication that each morph had been formed in a single, large mutation.

20. Wallace, "Malayan Papilionidae," pp. 10-11. In the first edition of the *Malay Archipelago*, Wallace repeated the story in the text of p. 140 rather than relegating it to the footnotes as he had in the "Malayan Papilionidae."

21. Wallace, "Malayan Papilionidae," p. 14.

22. Wallace, "Malayan Papilionidae," p. 10. Wallace wrote, "Dimorphism is thus seen to be a specialized result of variation, by which new physiological phenomena have been developed." For him, polymorphism was a means by which a single species could exploit several different habitats--those

of the different models.

23. A. R. Wallace, "Notes on Eastern Butterflies," *Transactions of the Entomological Society of London, 1869, pp. 277-288.* In the context of suggesting a redefinition of a species, *Diadema Anomala*, Wallace pointed out a case in which the *male is the much* duller in color of the two sexes. In this species, the male is a dull brown and the female a rich blue. He asserted that females are not dull in color because they are females, but because the dull coloration is protective for them. If they were protected by brilliance there is nothing to prohibit or inhibit that from happening in female as well as male organisms.

24. Wallace, "Malayan Papilionidae," p. 22.

25. Wallace, "Malayan Papilionidae," p. 17. Wallace asked, "Is not the a priori evidence in favour of the assumption of an identity of the causes that have produced such similar results? and have we not a right to call upon our opponents for some proofs of their own doctrine, and for an explanation of its difficulties, instead of their assuming that they are right, and laying upon us the burthen of disproof?"

26. J. C. F. Fryer, "An Investigation by Pedigree Breeding into the polymorphism of *Papilio polytes*," *Philosophical Transactions of the Royal Society, 204*:227-254, 1913-14. Fryer called the species *P. polytes* rather than *P. pammon* as Wallace did; it is the same species however. The male form is called *P. polytes* and the female form, which resembles the male, is called *cyrus* (non-mimetic). The female form, *romulus*, is mimetic, resembling *Papilio hector*. Another female form, *polytes*, also mimetic, resembling *P. aristolochi*. (Fryer's paper hereinafter referred to as "*Papilio Polytes*."

27. Fryer, "*Papilio Polytes*," p. 237. Fryer is dealing with factors for maleness and femaleness, a factor for mimicry and an inhibiting factor which prevents mimicry.

28. Fryer, "*Papilio Polytes*," p. 249. In the absence of selection, $q^2=pr$, when p=number of pare nominants, and 2q=number of heterozygotes and r=number of recessives.

29. Reginald Crundell Punnett, *Mimicry in Butterflies* (Cambridge University Press, 1915).

30. Punnett, *Mimicry in Butterflies*, p. 7.

31. Punnett, Mimicry in Butterflies, Chapter 7, pp. 75-92 and Chapter 8, pp. 93-103.

32. Punnett, Mimicry in Butterflies, p. 51.

33. Punnett, Mimicry in Butterflies, pp. 54-56.

34. Punnett, Mimicry in Butterflies, p. 57.

35. Punnett, Mimicry in Butterflies, p. 60.

36. Alfred Russel Wallace, Darwinism. An exposition of the Theory of Natural Selection with Some of its Applications (London: Macmillan, 1889). The edition used here is the second edition. Quotation on p. 180.

37. Alfred Russel Wallace, Contributions to the Theory of Natural Selection (New York: Macmillan, 1871) p. 76. Emphasis added to the quotation.

38. Wallace, Darwinism, p. 180. Of course, it was possible to disbelieve in the validity of Wallace's laws of mimicry and still be a Darwinian. The best case of this position is Fritz Müller's, who discovered a new form of mimicry precisely because he had doubts about Wallace's laws.

39. Wallace, "Malayan Papilionidae," p. 31.

40. Alfred Russel Wallace, "Mimicry and other Protective Resemblances among Animals," Westminster Review, July 1, 1867, pp. 1-43. Reprinted in Natural Selection with additions and corrections.

41. Edward Forbes, "On the Manifestation of Polarity in the Distribution of Organized Beings in Time," Edinburgh New Philosophical Journal 57:332-337, 1854. See, for example of the Quinary Classification System: William Swainson, On the Natural History and Classification of Birds (London: Longman, Rees, Orme, Brown, Green & Longman, 1836).

42. Alfred Russel Wallace, Contributions to a Theory of Natural Selection, p. 35.

43. Sweet peas had been studied by several Mendelians including Francis Galton, William Bateson, E. Saunders and R. Punnett. Primula also was studied by Charles Naudin and R. P. Gregory. See Elof A. Carlson, The Gene: A Critical History (Philadelphia: W. B. Saunders, 1966) pp. 42, 55; and Robert C. Olby, Origins of Mendelism (New York: Schocken Books,

1966) pp. 62, 77-78. See also the article by L. DeRuiter, "Some Remarks on Problems of the Ecology and Evolution of Mimicry," *Archives Neolandaises du Zoologie 13*, supplement 1958, pp. 351-368 and Hugh B. Cott, *Adaptive Coloration in Animals* (London: Methuen, 1940) p. 406.

44. Punnett, *Mimicry in Butterflies*, p. 91.

45. Punnett, *Mimicry in Butterflies*, p. 92.

46. Punnett, *Mimicry in Butterflies*, p. 96.

47. Punnett, *Mimicry in Butterflies*, p. 100.

48. Punnett, *Mimicry in Butterflies*, p. 153.

49. Punnett, *Mimicry in Butterflies*, p. 152.

50. Punnett, *Mimicry in Butterflies*, p. 152.

51. Ronald Fisher, "On Some Objections to Mimicry Theory: Statistical and Genetic," *Transactions of the Entomological Society of London 75*:269-278, 1927; and E. B. Ford, "The Genetic Basis of Adaptation," in *Evolution*, edited by G. R. deBeer (Oxford: Clarendon Press, 1938), pp. 43-55. Fisher advanced the idea in his article that Mendelian factors determine whether an organism will be mimetic or non-mimetic, but that the alternatives open to an organism may be modified in the course of evolutionary development. Punnett's mistake had been to assume that the Mendelian factors were static rather than dynamic. Although the options of mimetic or non-mimetic are like a yes or no answer, many changes in the character of mimicry may occur in time through selecting agents. Fisher described the phenomenon of stabilizing selection with respect to Batesian mimicry. He demonstrated that the number of individuals in a species would be increased by Batesian mimicry to two or more models. The stable ratio of non-mimetic to mimetic forms of *Papilio polytes* is, then, both evidence of selection and of the "dependence of selective advantage upon the actual ratio of the alternative types" (p. 276). When the selective advantage depends upon the relative frequency of different morphs of a polymorphic species, a stable ratio is often established and maintained by natural selection.

Fisher was answering critics of Darwinism in biometric and genetic terms. His work was reinforced by E. B. Ford's studies of the pigments found in butterfly scales. Ford found that there were several possible evolutionary pathways to the same visual result. Both *Papilio polytes* and its model P. hector have red patches on their hind wings. Ford found that the red pigments were very different chemically. The findings invalidated Punnett's position that mimicry arose by duplicate saltations in both mimic and model.

The controversy was reopened by papers in 1945 by Richard Goldschmidt in which he redefined "parallel mutation" to avoid Ford's criticism and oddly sided with Punnett. Arguments in Darwinism are very hard to finish. "Mimetic Polymorphism, a Controversial Chapter of Darwinism," Quarterly Review of Biology 20:147-164, 205-230, 1945.

52. Roland Trimen, "On some Remarkable Mimetic Analogies among African Butterflies," Linnean Society Transactions 26:497-523, 1870. (Hereinafter referred to as "Remarkable Mimetic Analogies.")

53. Trimen, "Remarkable Mimetic Analogies," p. 498.

54. Jean Alphonse Boisduval (1801-1879), Species Général des Lépidoptères, pp. 372-373. Quoted by R. Trimen as a heading for his paper. The Papilio Westermanni of Boisduval was later the Papilio Merope (female form Hippocoon) according to Trimen's usage. It is now Papilio dardanus (female form, Hippocoon).

55. Trimen, "Remarkable Mimetic Analogies," p. 497.

56. Trimen, "Remarkable Mimetic Analogies," p. 520.

57. Roland Trimen, "Observations on the Case of Papilio Merope, Auct.; with an account of the various known forms of that Butterfly," Transactions of the Entomological Society of London for 1874, pp. 137-153.

58. Trimen, "Observations on the Case of Papilio Merope..." p. 138.

59. Roland Trimen, "On some undescribed Species of South-African Butterflies, including a New Genus of Lycaenidae," Transactions of the Entomological Society of London for 1868, pp. 69-77 and "Aspects of Insect Life in South-Eastern Africa," Entomological Monthly Magazine 4: 212-223, 1868.

60. Trimen, "On some undescribed Species...," p. 75.

61. Trimen, "On some undescribed Species...," p. 75. The staff of the museum had placed a model specimen in the

sequence of _Papilio_ mimics. The _boisduvalianus_ is one of the forms of _Papilio Merope_ according to Trimen's nomenclature.

62. Trimen, "Aspects of Insect Life," pp. 220-221.

63. Trimen, "On some undescribed Species...," p. 69.

64. William Chapman Hewitson (1806-1878), was the owner of what was reputed to be the most complete private butterfly collection in the world. He bequeathed his collection to the National Collection in the British Museum at South Kensington. He was a member of the Zoological, Linnean, and Entomological Societies.

65. William Hewitson, letter press accompanying the five volumes of _Exotic Butterflies_ part 72, October 1869. Quotation taken from Trimen, "Observations on the Case of _Papilio Merope_...," p. 140.

66. Hewitson from Trimen, "Observations on the Case of _Papilio Merope_...," p. 140.

67. Hewitson from Trimen, "Observations on the Case of _Papilio Merope_...," p. 140.

68. Hewitson from Trimen, "Observations on the Case of _Papilio Merope_...," p. 141.

69. Hewitson from Trimen, "Observations on the Case of _Papilio Merope_...," p. 141.

70. J. P. Mansel Weale, "Notes on the Habits of _Papilio Merope_, with a description of its Larva and Pupa," _Transactions of the Entomological Society of London for 1874_, pp. 131-136; and William Hewitson, "Note on the Capture of _Papilio Antimachus_, &c.," _Entomologists' Monthly Magazine_ _11_:113, 1874.

71. Trimen, "Observations on the Case of _Papilio Merope_...," p. 141.

72. Hewitson, "Note on the Capture of _Papilio Antimachus_, &c.," _Entomologists' Monthly Magazine_ _11_:113, 1874.

73. Charles Darwin, _The Descent of Man and Selection in Relation to Sex_ (New York: Modern Library), p. 667. First published 1871. August Weismann, _The Evolution Theory_, translated by J. Arthur Thomson and Margaret R. Thomson (London: Edward Arnold, 1904) Volume 1, pp. 94-95 and Plate 1.

74. Weismann, Evolution Theory, I, pp. 91-118.

75. Weismann, Evolution Theory, I, p. 93.

76. Weismann, Evolution Theory, I, p. 95.

77. Guy A. K. Marshall (1871-1959), "Five Years Observations and Experiments (1896-1901) on the Bionomics of South African Insects chiefly directed to the investigation of mimicry and warning colours," with discussion by E. B. Poulton, Transactions of the Entomological Society of London for 1902, pp. 287-541.

78. Hnery Walter Bates, "Contributions to an Insect Fauna of the Amazon Valley. Lepidoptera: Heliconidae," Transactions of the Linnean Society of London 23:495-566, 1862. Bates wrote: "I never saw the flocks of slow-flying Heliconidae in the woods persecuted by birds or Dragon-flies, to which they would have been easy prey; nor, when at rest on leaves, did they appear to be molested by lizards or the predacious Flies of the family Asilidae, which were very often seen pouncing on Butterflies of other families. If they owe their flourishing existence to this cause, it would be intelligible why the Leptalidae, whose scanty number of individuals reveals a less protected condition should be disguised in their dress, and thus share their immunity" (p. 510).

79. Erich Haase, Untersuchungen über die Mimicry auf Grundlage einer naturlichen Systems der Papilioniden, Vol. 8, Bibliotheca Zoologica 1893 (Stuttgart: Erwin Nägele).

80. Weismann, Evolution Theory, Vol. I, page facing Plate I, between pages 112-113.

81. Edward B. Poulton, The Colours of Animals (New York: D. Appleton, 1890), pp. 181-182. Guy A. K. Marshall, "Five Years' Observations...," Transactions of the Entomological Society of London for 1902, pp. 328-337, contain the correction of Bates that predaceous insects do attract butterflies which are distasteful to birds. Page 337 contains the following statement: "A high degree of protection from the attacks of the generality of insect-eating animals will always be found to be compensated by the attacks of special enemies, and probably very largely by that of insect parasites. I brought forward this argument in 1890 (Colours of Animals, London, p. 181); and Haase, without attempting to meet it, made the crude assumptions which will now be discussed once and for all, by the numerous observations recorded below."

Poulton did not let it go at that; he mentioned Haase's error in "The Place of Mimicry in a Scheme of Defensive Coloration," p. 318, in his *Essays on Evolution* (Oxford: Clarendon Press, 1908). In that same essay he showed how Haase had led Weismann astray in the use of the term "immune" for the distasteful models, p. 375.

82. Some examples include: George F. Leigh, "Syn-epigonic series of *Papilio Cenea* (1902-3) and *Hypolimas missipus* (1904) together with observations on the life-history of the form. With notes by Prof. E. B. Poulton and an Appendix by Roland Trimen," *Transactions of the Entomological Society of London for 1904*, pp. 677-695. E. B. Poulton, "Mimetic Forms of *Papilio dardanus* (*merope*) and *Acraea johnstoni*," *Trans. Ent. Soc. 1906*, pp. 281-321; and E. B. Poulton, "Heredity in six families of *Papilio dardanus*, *Brown*, Subspecies *Cenea*, *Stoll*., bred at Durban, by Mr. G. F. Leigh," *Trans. Ent. Soc. 1908*, pp. 427-445.

83. Harry Eltringham, *African Mimetic Butterflies* (Oxford: Clarendon Press, 1910).

84. Eltringham, *African Mimetic Butterflies*, p. 86.

85. C. A. Clark and P. M. Sheppard, "Interactions between major genes and polygenes in the determination of the mimetic patterns of *Papilio dardanus*," *Evolution 17*: 404-413, 1963.

86. Wolfgang Wickler, *Mimicry in Plants and Animals* World University Library (New York: McGraw-Hill Book Co., 1968), pp. 26-33.

87. Fritz Müller [Johann Friedrich Theodore] (1822-1897), Ph.D. University of Griefwold in 1844; a student of many areas of natural history including botany, fertilization, hybridization and sterility problems. His most well-known works are *Für Darwin* and his work on mimicry. "The Darwinian Hypothesis supported by Observations on Crustacea," *Annals and Magazine of Natural History 15*:410-416, 1865.

88. Fritz Müller,"Einege Worte über Leptalis," *Jenaische Zeitschrift für Naturwissenschaft for 1876*:1-12.

89. Oscar Schmidt, *Descendenztheorie und Darwinismus*, 1873. English translation called *The Doctrine of Descent and Darwinism* (New York: D. Appleton & Co., 1875). See pp. 179-182.

90. Fritz Müller, "Notes on Brazilian Entomology," Transactions of the Entomological Society of London for 1878:211-223, pp. 222-223.

91. Fritz Müller, "Ituna und Thyridia. Ein merkwürdiges Beispiel von Mimicry bei Schmetterlingen," Kosmos 5:100-108, 1879. [Translated into English and communicated by Raphael Meldola as] "Ituna and Thyridia: a Remarkable Case of Mimicry in Butterflies," Proceedings of the Entomological Society of London for 1879:xx-xxix. See pp. xxiv-xxv.

92. In England Meldola had exhibited a collection of Indian insects which clearly demonstrated the effectiveness of the protection granted these protected genera; most of the other specimens had been destroyed by predaceous mites. Proceedings of the Entomological Society for 1877, p. xii.

93. Müller, "Ituna and Thyridia," Proceedings of the Entomological Society for 1879, p. xxvii. All quotations are from the English translation.

94. Müller, "Ituna and Thyridia," p. xxvii.

95. Müller, "Ituna and Thyridia," p. xxviii. The species involved in mutual mimicry probably approach one another through selection based on the advantage of mimicry, but they are already quite close in genetic affinity.

96. Müller, "Ituna and Thyridia," p. xxv. "The more closely related are the two species which resemble one another, the more alike they were ab initio, and the more uncertain in consequence is their second indication of mimicry; it becomes perfectly useless in cases where the nearest allies of the two species are without a common, peculiar, sharply defined form of marking and coloring. Coloenis Julia and Eueides Aliphera will again serve as examples. In the genus Coloenis, near the fiery-red Julia, there stands the green Dido, and the other species with still different colours and form of wing, whilst in the genus Eueides, the spotted Isabella and the Acraea-like Pavana stand near the fiery-red Aliphera."

97. Alfred Russel Wallace, "Dr. Fritz Müller on Some Difficult Cases of Mimicry," Nature 26:36-37, 1882.

98. Fritz Müller, "Bemerkenswerthe Fälle erworbener Aehnlichkeit bei Schmetterlingen," Kosmos X Band, V Jahgang, 1881-82, pp. 257-267.

99. [Column headings] "Difficult Cases of Mimicry," Nature 26, p. 105; 27, p. 481 and "Protection by Mimicry--a Problem in Mathematical Zoology," Nature 29:405-406, 1884.

100. Thomas Blakiston and Thomas Alexander, "Protection by Mimicry--A Problem in Mathematical Zoology," Nature 29:405-406, 1884.

101. Blakiston and Alexander, "Protection by Mimicry," 1884, p. 406.

102. William L. Distant, "Difficult Cases of Mimicry," Nature 26:105, 1882.

103. Distant referred to studies by Doublas Spalding on domestic turkeys.

104. William Lucas Distant (1845-1922), Rhopalocera Malayana: A Description of the Butterflies of the Malay Peninsula (London: printed by W. L. Distant, 1882-86), p. 33. Distant doubted that "mimicry" was the right term for cases like Müller's. He did not dispute Batesian mimicry, but only expressed doubt that the common resemblances Müller found were the same thing as true mimicry. The need for a means to distinguish the terms of Batesian from that of Müllerian mimicry was only supplied in 1897 by E. B. Poulton. Poulton, with the aid of Arthur Sidgwick, designated Batesian mimicry Pseudaposematic coloration or "false warning colors." Proceedings of the Entomological Society of London for 1897, p. xxix.

105. Raphael Meldola (1849-1915) F. R. S. known for works in chemistry as well as in entomology. He served in many capacities as a member of the Entomological Society of London.

106. Raphael Meldola, "Mimicry Between Butterflies of Protected Genera," Annals and Magazine of Natural History series 5, 10:417-425, 1882. Quotation from p. 419.

107. Meldola, "Mimicry Between Butterflies of Protected Genera," pp. 420-421.

108. Meldola, "Mimicry Between Butterflies of Protected Genera," p. 420.

109. Distant associated the Darwinians with a surrender to theoretical partisanship and to the abandonment of healthy scepticism. He opposed what he thought was an incautious

extension of the evolutionary theory and he was adversely disposed to the pretense that Darwinian evolution explained everything.

110. William L. Distant, "The Theory of Mimicry and Mimicking Theories," Annals and Magazine of Natural History, series 5, 11:43-48, 1883

111. John Jenner Weir (1822-1894), Fellow of the Linnean Society, and of the Zoological Society. Vice President of the Entomological Society 1875, 1885-86. Member of the Council for many years.

Jenner Weir began his tests in the summer of 1868 at the suggestion of Alfred Russel Wallace. He published his results in two papers. The species tested were common ones such as the robin, the Bullfinch, Chaffinch, Thrush, Crossbill, Tree Pipit, and the Reed Bunting, as well as the Siskin and the Redpoll to a lesser degree. "On Insects and Insectivorous Birds; and especially on the Relation between the Colour and the Edibility of Lepidoptera and their Larvae," Transactions of the Entomological Society of London 1869, pp. 21-26; and "Further Observations on the Relation between the Colour and the Edibility of Lepidoptera and their Larvae," Transactions of the Royal Entomological Society of London 1870, pp. 337-339. A. G. Butler studied the responses of green lizards (*Lacerta viridis*) to various caterpillars and published his results following Weir's first paper. "Remarks upon certain Caterpillars, &c., which are unpalatable to their enemies," Transactions of the Entomological Society 1869, pp. 27-29.

112. Müller, "*Ituna* and *Thyridia*," p. xxvii.

113. Jenner Weir was apparently first to respond following the reading of Meldola's translation of "*Ituna* and *Thyridia*" by Fritz Müller. His opinions are to be found in the Proceedings of the Entomological Society 1879, p. xxviii. "Mr. Jenner Weir stated that from the numerous experiments which he had made on the subject of larvae which are eaten or rejected, he had always been profoundly impressed with the utter disregard paid by birds to larvae which were inedible. He had never but once seen a distasteful larva even examined by a bird. When, day by day, he had thrown into his aviary various larvae, those which were edible were eaten immediatly; those which were inedible were no more noticed than if a pebble had been thrown before the birds. It was Mr. Weir's opinion that the experience of birds in this respect had become hereditary in the species, and was not the result of the experience of individual birds, but

was rather to be regarded as an act of 'unconscious cerebration.'"

114. [Bates's participation in the discussion of "Ituna and Thyridia"] Proceedings of the Entomological Society of London 1879, pp. xxviii-xxix.

115. [Bates's participation in the discussion of "Ituna and Thyridia"] Proceedings of the Entomological Society of London 1879, p. xxix.

116. [Bates's participation in the discussion of "Ituna and Thyridia"] Proceedings of the Entomological Society of London 1879, p. xxix. E. B. Poulton affirms that Bates never accepted Müller's hypothesis. Poulton, Essays on Evolution (Oxford: Clarendon Press, 1908), p. 223.

117. Alpheus Spring Packard (1839-1905). Studied with Louis Agassiz; was co-founder of the American Naturalist and editor for twenty years; one of the first students of insect embryology and introduced the study of comparative embryology. "The Origin of the Markings of Organisms (Poecilogenesis) Due to the Physical rather than to the Biological Environment; with criticisms of the Bates-Müller Hypothesis," Proceedings of the American Philosophical Society 43:393-450, 1904. See pp. 416-420.

118. Packard, "The Origin of the Markings...," p. 416.

119. Packard, "The Origin of the Markings...," p. 420.

120. George F. Hampson (1860-1936), "Protective and Pseudo-Mimicry," Nature 57, February 17, 1898.

121. Hampson, "Protective and Pseudo-Mimicry," Feb. 17, 1898.

122. "Proceedings of the South London Entomological and Natural History Society, 1907-1908," The Entomologist's Record and Journal of Variation 20:189, 1908.

123. "Proceedings So. London Ent. and Nat. Hist. Soc.," p. 190.

124. Guy A. K. Marshall, "On Diaposematism with Reference to Some Limitations of the Müllerian Hypothesis of Mimicry," Transactions of the Entomological Society of London 1908:93-142.

125. "Proceedings So. London Ent. and Nat. Hist. Soc." p. 190.

126. Frederick A. Dixey, "On the Phylogeny of the Pierinae, as Illustrated by their Wing-Markings and Geographical Distribution," Transactions of the Entomological Society of London 1894:249-334. "On the Relation of Mimetic Patterns to the Original Form," Trans. Ent. Soc. Lond. 1896:65-77. "Mimetic Attraction," Trans. Ent. Soc. London 1897:317-332. For the discussion of Dixey's 1897 paper see: Proceedings of the Entomological Society of London 1897:xx-xliv.

127. E. B. Poulton, [Discussion], Proceedings of the Entomological Society of London 1897:xxix. See also: Poulton, Essays on Evolution, 1908. Chapter 10. "The Place of Mimicry in a Scheme of Defensive Coloration," pp. 293-382, especially 342-356.

128. Poulton, Essays on Evolution, p. 212.

129. Andrew Murray (1812-1878). Fellow of the Royal Society of Edinburgh, Linnean Society; assistant secretary of the Royal Horticultural Society and its scientific director after 1877. His most important work is The Geographical Distribution of Mammals (London, 1866). His special interest in plants was in the Coniferae and in invertebrates the Coleoptera. "On Mr. Darwin's Theory of the Origin of Species," Proceedings of the Royal Society of Edinburgh 4:274-292, 1862. "Reply to Mr. Wallace's Theory of Birds' Nests," Journal of Travel I:137-145, 1869.

130. A. Murray, "Mimicry and Hybridisation," Nature 3:154, 1870.

131. Murray is aware of cases in which the results of reciprocal crosses are very different. Murray is most familiar with examples from botany. He believed that sexual dimorphism in butterflies was due to a combination of environmental factors and hereditary results of hybridisation.

132. Murray, "Mimicry and Hybridisation," Nature 3:154, 1870.

133. Letters by Arthur G. Butler, Alfred Russel Wallace in Nature 3:165-66, 1870.

134. Two works which discuss the Darwinian problems with the utility of incipient structures and the relationship

of evolution to hybridization: see Peter J. Vorzimmer, *Charles Darwin: The Years of Controversy* (Philadelphia: Temple University Press, 1970); and Michael T. Ghiselin, *The Triumph of the Darwinian Method* (Berkeley: University of California Press, 1969).

135. Alfred W. Bennett, "The Theory of Natural Selection from a Mathematical Point of View," *Nature* *3*:30-33, 1870.

136. Bennett was relying upon the work of Joseph John Murphy, *Habit and Intelligence, in their connection with the Laws of Matter and Force*, 2 vols. (London: Macmillan, 1869). The reference here seems to be to Volume I, pp. 323-325.

137. Bennett, "Theory of Natural Selection...," p. 32.

138. Bennett, "Theory of Natural Selection...," p. 33.

139. Bennett, "Theory of Natural Selection...," p. 33.

140. Bennett, "Theory of Natural Selection...," p. 30. Bennett has also been influenced by another anti-Darwinian, the Duke of Argyll, (George Douglas Campbell), *The Reign of Law*, 1867.

141. Alpheus Packard, "The Markings of Organisms (Poecilogenesis) Due to the Physical rather than to the Biological Environment with Criticism of the Bates-Müller Hypothesis," *Proceedings of the American Philosophical Society* *43*:393-450, 1904.

142. Packard, "Markings of Organisms," pp. 400-401. He used both published material and personal communications. Some of the people quoted were: Dr. J. A. Allen, S. H. Scudder, Mr. F. P. Drowne, Mrs. Annie T. Slosson, Prof. Cockerell, Prof. W. M. Wheeler, Mr. H. H. Hill, Miss Geraldine Street, Mr. C. V. Riley, etc.

143. Packard, "Markings of Organisms," p. 417.

144. E. B. Poulton, "The Experimental Proof of the Protective Value of Colour and Markings in Insects in Reference to their Vertebrate Enemies," *Proceedings of the Zoological Society of London*, *1887*:191-274.

145. Packard, "Markings of Organisms," pp. 421, 429.

146. Packard, "Markings of Organisms," p. 405.

Packard wrote of the Monarch butterfly, "On cutting out a bit of the underside of the middle of the abdomen I tasted it and perceived a very slight taste not like any substance known to me, and certainly not resembling that of laudanum. On removing a larger piece and pressing the abdomen the odor became more distinct, and one of my family at once detected it, although at first before it was wounded she could perceive no odor. I tasted two pieces cut from the middle of the abdomen, but the taste was hardly perceptible and not unpleasant, neither bitter, acid, or in any way pungent."

147. Packard, "Markings of Organisms," pp. 421, 442-446.

148. Packard, "Markings of Organisms," p. 442.

149. Packard, "Markings of Organisms," p. 443. Packard referred to works by Thayer, for example: "The Law which underlies protective coloration," The Auk 13:124-129, 1896; and "Further Remarks on the Law which underlies protective coloration," Auk 13:318-320, 1896; and "Protective Coloration in its relation to mimicry, common warning colours, and sexual selection," Transactions of the Royal Entomological Society of London 1903:553.

150. Thirty-five years elapsed between Walsh and Riley's paper on the Monarch (Danais archippus) and the mimic (Nymphalis disippus) and Packard's paper. Packard called the model the Milkweed Butterfly (Anosia plexippus). The naming of the species is in flux during the period. It is now called Danaus plexippus. Its mimic is called by Packard Basilarchia archippus (disippus); it is presently called Limenitis archippus.

151. Frederick A. Dixey, "On the Phylogeny of the Pierinae, as illustrated by their Wing-Markings and Geographical Distribution," Trans. Ent. Soc. London 1894:249-334. "On the Relation of Mimetic Patterns to the Original Form," Trans. Ent. Soc. London 1896:65-77. "Mimetic Attraction," Trans. Ent. Soc. Lond. 1897:317-332. See also E. B. Poulton, Essays on Evolution, pp. 213, 223, 328, 342-348. Poulton wrote: "During the past thirteen years the Müllerian theory has gradually absorbed more and more of the ground that was formerly believed to be covered by the theory of Bates--an advance due in chief part to the researches of Dr. F. A. Dixey." (p. 223)

152. A. Franklin Shull, "The Needs of the Mimicry

Theory," *Science* *85*:496-498, 1937. Quotation, p. 497.

153. Ronald Fisher, "On Some Objections to Mimicry Theory: Statistical and Genetic," *Trans. Ent. Soc. London* *75*:269-278, 1927; and *The Genetical Theory of Natural Selection* (Oxford: Clarendon Press, 1930). G. D. Hale Carpenter, "Experiments on the Relative Edibility of Insects, with special reference to their Coloration," *Trans. Ent. Soc. London* *1921*:1-105. "Further Evidence that Birds do attack Butterflies," *Proc. Zool. Soc. London* *A*,*3*:223-247, 1937. G. D. Hale Carpenter and E. B. Ford, *Mimicry* (London: Methuen, 1933). E. B. Ford, "Problems of Heredity in the Lepidoptera," *Biological Review* *12*:461-503, 1937.

154. W. L. McAtee, "The Experimental Method of testing the Efficiency of Warning and Cryptic Coloration in Protecting Animals from their Enemies," *Proceedings of the Academy of Natural Sciences Philadelphia*, *1912*, pp. 281-364; and *Effectiveness in Nature of the so-called Protective Adaptations in the Animal Kingdom, chiefly as illustrated by the Food-habits of Nearctic Birds* (Smithsonian Miscellany Collection, Washington 85, (7), 1932). A good discussion and criticism of McAtee's methods is found in Hugh Cott, *Adaptive Coloration in Animals* (London: Methuen, 1940), pp. 168-171.

155. Frank Finn (1868-1932), "Contributions to the Theory of Warning Colors and Mimicry. No. IV Summary and Conclusions," *Journal of the Asiatic Society of Bengal* *66*: 613-668, 1897.

156. "Moral Reflections at the Natural History Museum," *Punch*, May 2, 1906, pp. 312-313. Pouton quoted part of this article in *Essays on Evolution*, pp. 213-215.

157. *Punch*, May 2, 1906, pp. 312-313.

158. Alvar Ellegard, *Darwin and the General Reader. The Reception of Darwin's Theory of Evolution in the British Periodical Press, 1859-1872* (Goteborg: Gothenburg Studies in English 8, 1958). Ellegard demonstrates that Darwinian critics eagerly associated Darwinian views of descent with the discredited and refuted theories of Lamarck, pp. 269-270. Darwin's and Lamarck's views are consistently conflated in the magazine literature Ellegard surveyed.

159. W. L. Distant, "The Theory of Mimicry and Mimicking Theories," *Annals and Magazine of Natural History* *11*:43-48, 1883. Quotation, p. 46.

160. William Kirby and William Spence, *An Introduction to Entomology: or, Elements of the Natural History of Insects*, 6th edition (Philadelphia: Lea and Blanchard, 1846). See Letter 21, "Means by which Insects defend themselves." The fly, *Volucella*, resembles the bee of the genus *Bombus*, p. 442.

Chapter IV

EVOLUTION AT THE ENTOMOLOGICAL SOCIETY OF LONDON, 1860-1875

Seventeen members of the Entomological Society of London who were original members when the society was founded in 1833 were still listed as members in 1858 when the Darwinian controversy was about to begin.[1] The name Charles Darwin appears on the list of October 1858 and of 1833; he remained a member of the Entomological Society for the entire forty-nine year period between the founding of the society and his death in 1882, although he rarely came to meetings in London.[2] His ideas, however, had very great importance for the entomologists in the society and frequently dominated the discussions. Two members who were important in bringing Darwin's ideas before the Society, Alfred Russel Wallace and Henry Walter Bates, were travelling in distant countries in 1858-59, when evolutionary ideas became an important topic in British science. In 1858 Wallace's and Bates's names were listed as corresponding members, both having been elected to membership in 1857.

The Entomological Society of London is a particularly interesting group to study in connection with the reception of the Darwinian theory of evolution. Access to membership in the society had been made easy. The constitution of the group, adopted at the founding in 1833, permitted a member to be elected by a two-thirds vote of the members after the candidate is proposed for membership by three currently active members. Because it was possible for almost any aspiring entomologist to become a member, young men and amateurs like Wallace and Bates were members of the society on equal terms with the entomologists from the British Museum.[3] Numerically the society was dominated by amateurs; only about thirty percent of the members listed an college degree or membership in another scientific or learned society after his name in the 1858 list of members.

The period of the first debates about Darwinism was also a time in which other issues were being debated in the group which affected the reception of evolutionary theory. The age was one of transition; entomologists in the period were concerned to improve the status of their discipline and to enhance its recognition as a science. The addresses of presidents of the society show a concern for the improved image of the discipline. The articles published by the Society were mostly by authors who came from the professional or near-professional ranks of the members. The _Proceedings of the Entomological Society_ contained many papers given; for example, major revisions of taxonomic groups and papers describing new species. The _Proceedings_ did not very often carry "notes and observations" by amateur naturalists. There were other magazine-style weeklies and monthly publications in which news of rare insects and remarks on insect behavior were published, for instance, in the *Entomologist's Monthly Magazine* and in the *Entomologist's Weekly Intelligencer*. The *Proceedings of the Entomological Society* reflected the hopes of the members who aspired to high status in science for themselves and for their studies. The concern in the society for its image and stature as a scientific discipline intersected with its concern for adopting good methodology of scientific practice and theoretical working principles.

The debates at the society concerning evolution were introduced at a time when members were concerned that entomological study be perceived as a serious discipline, not dominated by either mere collectors or by enthusiasts. For a long time, entomology had been connected in the minds of many with amateur natural history. Also, it had become one of the possible forms of investment for wealthy people; exotic species won very high prices. The leaders of the Entomological Society were anxious that their subject be dissociated from both the commercial interests of a leisured class and the casual pursuits of hobbyists. Concern for the future of entomology precedes the publication of the *Origin of Species*, but became more overt and intense after 1859 because of the new theory Darwin and Wallace had proposed.

Appropriate Attitudes in Entomology

The process of collecting and classifying specimens had begun to appear shallow. A paper touching on the future of entomological work by Henry Tibbats Stainton was read at the Entomological Society in 1856. His paper, "How may the Onward Progress of the Sutyd of Entomology be Best furthered?" laid stress on the importance of the attitude with which entomologists approached their work.[4] Stainton was concerned with the superficial nature of entomological study. He made a distinction between "information" which may be derived from the work of others and "knowledge" which is only to be achieved by one's own study of objects themselves and is to be obtained only by strenuous effort to penetrate the mysteries of one branch of study. Understandably, Stainton encouraged his colleagues in the swift pursuit of "knowledge" and encouraged their dissatisfaction with mere "information" which might be borrowed from the efforts of others.

Entomological studies were to be coupled with the pursuit of the virtue of unselfishness. In a metaphor reminiscent of the Proverb of Solomon to "go to the ant" for lessons in industry, Stainton directed entomology students to the bee for demonstrations of unselfishness.

> The bee rifles the flower of its honey not for its own immediate pleasure and enjoyment, but in order that it may be carried home and added to the common store for the future of the community; the bee swallows the honey and

> afterwards regurgitates it into a cell prepared for its reception; so it must be with the scientific student,--he must probe science to her innermost recesses, (he must remember that the bee buries herself in the corolla of some tubular flower,) and having acquired an amount of knowledge by actual investigation of objects themselves, he must arrange and classify the knowledge so obtained, and then, having thoroughly digested it, he must reproduce it in a simple, intelligible form so that those who have not time to pursue for themselves the peculiar branch of study to which he is devoted, may yet be able to derive advantage from his labours; just as the bees which remain occupied in the hive are benefited by the honey collected by those which go abroad.[5]

Although "information" is a desideratum of science, the scientists themselves ought not to be satisfied with it. Entomologists must guard themselves against the selfishness of private knowledge or of its circulation within only a small clique. Stainton wanted to emphasize that it is very important what attitude is developed among entomologists. Attitude will influence both the factual content and the theoretical structure of the developing science. Selfish attitudes will restrict the flow of facts from observers to others. Overemphasis on priority and a lack of humility had been too prevalent in the discipline. Arrogant attitudes in entomologists had caused some to publish premature attempts at wide-ranging theories in a time when collection of facts was still wanted.

Stainton's paper was an appeal for restraint among entomologists, not a radical Baconian tract. He felt, however, that entomologists' time was best spent in the acquisition of specific knowledge and in its publication. A measure of responsible generalization was permissible, even desirable, as long as the general conclusions stayed very close to "knowledge." Stainton's main complaint was about the poor methodology and attitudes which he detected in the entomological discipline. Other observers of science at the time found other, related, areas of concern for the future of entomology.

Darwin, in large measure, rejected the conservative, Baconian methodology with its emphasis on unadorned observation. The taxonomists among naturalists were the strongest proponents of an allegedly untheoretical approach to natural history. Darwin believed that the result of that philosophy had been that taxonomists named species without properly describing them.[6] Not enough thought about an organism preceded the rendering of name. A second and related problem had been perpetrated on science by taxonomists by the annexation of the name of the first describer to the name of the species. The result had been what Darwin called "species mongering." Taxonomists were motivated to name rather than to contemplate. The taxonomist's reputation depended upon the quantity of his work rather than on its quality. The more species he named, the greater his notariety.

Darwin found the state of taxonomy deplorable and said so in a letter to Hooker:

> I do not think more credit is due to a man for defining a species than to a carpenter for making a box. But I am foolish and rabid against species-mongers, or rather against their vanity; it is useful and necessary work which must be done; but they act as if they had actually made the species, and it was their own property.[7]

Stainton, as well as Darwin, held that entomologists had been seduced away from their legitimate purpose by the pursuit of reputation through the proliferation of species names without identities. By 1853, Darwin had begun to think that there was a kind of psychological incompatibility which made the co-existence of systematic and theoretical efforts in a single mind extremely difficult.

Having served more than an apprenticeship as a systematist in his study of the Cirripedia, Darwin moved his allegiance to the generalizers. Darwin did not feel that he himself suffered from the "miserable and degrading passion of mere species naming."[8] Taxonomy was more of a burden to be borne than an occupation which gave Darwin pleasure. The effort seemed to him fraught with frustration.

> After describing a set of forms as distinct species, tearing up my MS, and making them one species, tearing that up and making them separate, and then making them one again (which has happened to me), I have gnashed my teeth, cursed species, and asked what sin I had committed to be so punished.[9]

Darwin's views of taxonomy were based upon his long study of the crustaceans, the barnacles. He had asked himself the question of whether belief in the permanence of species affected the naming of species, and had decided that it had but little effect.[10] Although the actual practice of taxonomy was not much changed by belief or disbelief in evolution, taxonomists were seemingly barred from appreciating general ideas by their attention to particular data. Darwin wrote: "I really suspect there is something absolutely opposed to each other and hostile in the two frames of mind required for systematizing and reasoning on large collections of facts."[11] Entomology had long been attractive to those men whose minds were marked by their adherence to induction.

A review of the papers given at the Entomological Society in this period shows that some of the changes Stainton and Darwin wished for were slowly occurring. Bates and Wallace were each to have an important place in producing the kind of first-hand knowledge that Stainton had called for, accompanied by generalizations on the data that Darwin desired.

Letters from Abroad

Both Bates and Wallace sent letters to be read to members of the Society during their years of travel in South America and in the Malay Archipelago. The letters were full of first-hand "knowledge," in Stainton's sense of the word. The young travellers were in high repute among taxonomists because of the bountiful harvest in new species which they were sending to England. Their letters contained useful references to geographical variation which were welcomed by workers limited to museum work. It had been widely acknowledged that work was needed on geographical variation and distribution in order to properly relate species in higher taxonomic groups. Mapping the distributions of species and connecting data on geographical variation involved a different understanding of the nature of species than taxonomic work required. Wallace especially was concerned with placing data from geographical distribution studies in relation to the larger subject of zoogeography. In the nineteenth century, many questions about the causes, effects, and relations of the geographical distribution of animals were still unsettled. This uncertainty about the meaning of geographical distribution of animals made it particularly attractive as a subject with which to begin to build a new style of biological research.

Work like that being done by Wallace and Bates on the geographical distributions of insect species was not like the typical publications in the Proceedings of the Entomological Society. It tended to challenge the traditional definitions and understanding of species. Wallace wrote a little four-paragraph paper giving his definition of the term "species" as he had come to understand it after a long study of variation in populations of closely related animals.[12] He had decided that species must be given separate names if they were differentiated by stable, constant characters. Phenomena of complex geographical variations in single species and the serial replacement of species at the same latitude had tended to destroy Wallace's belief that taxonomy based on types had much utility. He had sought for different more natural criteria for making species designations. Some way to distinguish a variety or sub-species from a species needed to be devised. Wallace solved the problem by calling a group a separate species if it had constant characters separating it from near relatives.

Neither Wallace nor Bates questioned the validity of doing taxonomy; what they doubted was the validity of the belief that species are "absolute independent creations" while varieties are not. Wallace wrote:

> If an amount of permanent difference, represented by any number up to 10, may be produced by the ordinary course of nature, it is surely most illogical to suppose, and very hard to believe, that an amount of difference represented by 11 required a special act of creation to call it into existence.[13]

Bates was a little more discreet, but no less definite.

The Entomological Society members must have been surprised to hear Bates's "Notes on South American Butterflies" read on September 7, 1857.[14] In the paper he admitted that the Heliconidae appeared to him to be "mostly a modern creation, the species very susceptible of change in conjunction with the least modification of local circumstance." He finished his sentence, "But these theoretical notions, I suppose you do not care about."[15] It is likely that some members cared that the correspondent was a disbeliever in the permanence of species. It would be interesting to know if they also thought it appropriate for him to mention his doubts in a paper to be published in their society's Transactions. Neither Bates's hinted belief in changing species nor Wallace's law, that "every species has come into existence coincident in space and time with a pre-existing closely allied species," generated much public comment at the Entomological Society. In many ways, what Bates and Wallace were doing in their work on geographical distribution was out of the main-stream of interests of the society. The majority of entomologists were concerned with taxonomy, not with geography. Even the Wallace-Darwin papers of 1858 made little impression on the Entomological Society, in contrast to the ornithologists Alfred Newton and Canon H. B. Tristram.

It was really only after the publication of the *Origin of Species* that evolutionary theory and its relevance to insect taxonomy began to be discussed. Bates was the first member to make reference to Darwin's *Origin of Species*. Curiously, it was not in reference to Darwin's main thesis but a criticism of a statement by Darwin that the tropics were greatly affected by the last glacial age.[16] Since Bates was already a Darwinian, it seems odd that he did not make sure that his first mention of the Origin of Species before an audience probably hostile to evolution could be complimentary.

The Origin of Species at the Entomological Society

After having ignored the Darwinian theory in his Presidential Address for 1860, John W. Douglas felt he must make some mention of the new theory of natural selection in his address summarizing the events of 1861.[17] The context in which he mentioned the subject of natural selection was peculiar. He recalled that during the previous summer of 1861 all insects had been unusually scarce; the cold, wet weather of 1860 had killed many. Certain bees had been particularly depleted by the bad climatic conditions. Douglas relegated selection to a negative role, but doubted that even the strongest individuals would survive severer climatic conditions. Selection is prevented, Douglas thought, by extermination when the environmental conditions are hostile. Douglas concluded:

> It is not difficult to see how an exceptional season may not only render some species of insects rare for a time, but may altogether extinguish them in a locality or country, and that thus species may become extinct by a cause which overwhelms all the individuals in one common ruin, and leaves no one of them with a

> variation of constitution or modification of structure wherewith to begin a new species on the principle of "selection" with which we are so familiar.[18]

Darwin had carefully discussed the relation of natural selection to climatic changes and to extinction in several places in the *Origin of Species*.[19] Of course, if a species is utterly extinguished there is no selection, but Douglas's case of bees did not involve total extinction but temporary depletion. He stated in his address that about three-quarters of the normal population of solitary bees had been killed. The survival of hardy individuals was just the kind of evidence Darwin might have used to illustrate the positive role of natural selection. Selection not only works to eliminate ill-adapted forms, but to preserve "favored races in the struggle for life."

The Darwinian theory as a whole and Darwin's book, the *Origin of Species*, were both first discussed in a climate of hostile criticism at the Entomological Society. Papers published by the Society did not customarily discuss theories as a matter of their principal emphasis; and even in less formal addresses, it is obvious that the presidents would have preferred to recount the year's events without getting involved in controversial topics. It had been avoidable in the address for 1859 because of the death of William Spence, which required considerable time to memorialize, by enumerating his many contributions to entomology as a science and to the Entomological Society as an institution.[20] Douglas had avoided the topic in 1860 and in 1861 he mentioned the "principle of selection" without a direct reference to Darwin, Wallace or the Origin of Species. Nevertheless, the discussions of exhibits and the papers show an escalating interest in the relationship between Darwinian evolution and various aspects of entomology.

The question of the evolutionary origin of species was discussed explicitly in January of 1863 by an entomologist with an unlikely name for an evolutionist, William F. Kirby.[21] His paper had been written soon after the publication of his book, Manual of European Butterflies,[22] which was arranged so as to emphasize the geographical distribution of species, an approach to the study of insects he believed had been greatly neglected among entomologists. The paper, "On the Geographical Distribution of European Rhopalocera," is a good index of the influence of geographical study on the definitions of "species" and "variety." People who understood geographical distributions of species have as a rule a better chance of understanding the concepts of evolutionary change and of natural selection than workers in other sectors of natural history. Kirby had become impressed by the diversity of different kinds of variation found in nature. His paper presented his findings on the categories of variation which are very much like the six categories Wallace listed in his paper on the Malayan Papilionidae. Kirby distinguished individual variation from permanent varieties and from sexual dimorphism. He then compared the terms "sub-species" with the term "species." "A sub-species," W. F. Kirby stated, "is a permanent variety linked to its nearest ally by intermediate forms, but which entirely replaces the type in the localities in which it occurs."[23] The definition was straightforward and not subject to controversy had he concluded there. However, he continued, "A sub-species is not a true species, but under favourable circumstances it will become so in time, as the varieties intermediate between itself and the parent or cognate species die out."[24] Kirby was aware that variation, isolation, and extinction of intermediate forms leads to the evolution of new species.

Kirby had consulted faunal catalogues of butterflies for America and for the island of Corsica to compare with the European faunae. Finding the butterflies both of America and of Corsica different from European forms at the level of species, he concluded:

> If geologists can tell us how long Corsica has been separated from the main land, we shall have some most valuable data by which to calculate the length of time that the formation of a species requires, for a large number of Corsican insects are already good species, and many more have become sub-species, which is the last step towards the formation of a new species... In conclusion, I do not think any one present,

> however much he may differ from the opinions expressed in this paper, will regret my bringing prominently forward Mr. Darwin's theory of the Origin of Species, for it is the great scientific question of the day, and ought to be freely discussed here as well as in other societies; and my own opinion is, that, in a great measure, by the study of geographical distribution in its widest sense, the theory must stand or fall.[25]

It is not possible to tell if Kirby was able to catalyze the kind of "free discussion" he sought from the published minutes. The Transactions record simply, "Mr. W. F. Kirby read a paper 'On the Geographical Distribution of European Rhopalocera,' in which the author expressed himself as an adherent to the Darwinian theory of the origin and development of species."[26] It may have been that Kirby had underestimated the members' objections to his bringing "prominently forward" Darwin's theory. The names of those reading papers at that January meeting suggests that Kirby was, very probably, the only Darwinian in the hall. It also appears that Kirby's paper was the last to be read on what must have been a very long evening.

Members of the Society Confront Darwinism, 1860-1865

The intensity and frequency of discussion of Darwinian evolution increased when Wallace and Bates began to participate in the meetings. A quickened pace in treatment of evolutionary topics of several kinds is to be found from

1864 on. Batesian mimicry or the utility of other types of protective coloration was mentioned on a few occasions after Bates's paper "Contributions to an Insect Fauna of the Amazon Valley: Lepidoptera: Heliconidae" appeared in the Linnean Society Proceedings.[27]

Bates's memoir on mimicry, published in 1862, required comment from the President in his annual address in January of 1863. It was far too important a publication to pass unnoticed. The President, Frederick Smith, praised Bates's efforts and observations in the "'Metropolis' of the insect world." He recommended the paper to any who had not yet read it. He believed they would find its deductions "equally startling and conclusive." Smith may not have agreed with Bates, but he was not one to ignore an interesting study.[28]

In the year 1864, Bates's Naturalist on the River Amazons required comment from Smith, who was serving the second year of his presidency.[29] Smith heartily recommended the book and read excerpts to the audience. He noted that the "true science of Entomology teems in every chapter," without saying that Bates took great care to discuss geographical speciation as it is understood by advocates of Darwinian evolution.

Two months later, Frederick Smith presented an Anti-Darwinian paper to the April meeting of the Entomological Society, supporting the argument from design in the shape of bees' cells.[30] The subject of the hexagonal shape of bee cells had been used before by Rev. Samuel Haughton to dispute Darwinian evolution in general and natural selection in particular. Wallace had written an article as a rejoinder.[31] Smith's article, "On the Construction of Hexagonal Cells by Bees and Wasps," was different from Haughton's in its content. He tried to show that the bees were not building hexagonal cells to save space, or wax, or work: "No, the hexagonal form of cell is the plan laid down by the great Architect, and the bees are the builders that carry out His designs."[32] The Designer had implanted an instinct for hexagon building which the insects execute. The discussion which followed the reading of the paper shows that Smith was nearly as alone in his arguments as Bates and Wallace were in theirs. It was not that the opposition favored evolutionary explanation, but there was complaint about bringing "intention" into the discussion.

Edward Newman asked several rather impertinent questions. They were recorded in the minutes:

> Mr. Newman enquired to what extent Mr. Smith would carry his principle of intention: Would he apply it equally to the case of the larva cells of <u>Microgaster alveareus</u>? Was it applicable to the aggregation of lenses in the eye of insects? or could it afford any explanation of the hexagonal form of pillars of basalt?[33]

Mr. Smith's reply was not recorded. It was no longer adequate to refer an amazing and unexplained fact to design by the "great Architect." The fellows were willing to discuss whether the hexagons were constructed in circular shapes rendered hexagonal by being closely packed together or whether they were initially constructed as hexagons. The intrusion of teleology may have suggested to many of them the errors and excesses of an earlier time that was too content with theological interpretations of natural phenomena. If one was to be satisfied that the great Architect established the hexagonal shape of bee cells, one could also say that basalt crystals were formed at His behest as well.

From the spring of 1864 onward, the discussions of Darwinian evolution became more frequent and the published minutes make it appear to have been easier to get discussions started on evolutionary topics and that once begun the debates were freer. In the years 1864-66, Francis P. Pascoe was Society president. He was a believer in evolution but not in natural selection as its major mechanism. His reservations about Darwinism do not appear to have made him reticent to introduce exhibitions which would lend themselves to Darwinian analysis. It would seem that Pascoe was a person who did not shrink from the expression of diverse opinions on scientific matters. He had nominated men of three very distinct opinions to serve with him as Vice-Presidents: H. T. Stainton, A. R. Wallace, and Rev. Hamlet Clark.

Protective adaptations, particularly of coloration, were discussed in April, May, June, and July of 1864. On two occasions, letters were read from foreign correspondents relating to protective coloration. Bates read a letter from Roland Trimen in South Africa relating the story of a newly found spider that precisely resembled a flower.[34] In another meeting Bates read a letter from Benjamin Walsh of Rock Island, Illinois, on the mimetic resemblance of the *Danais plexippus* and the *Limenitis disippus*.[35] The letters must have served to underscore the wide occurrence of Bates's mimetic analogies. In the discussions of the exhibits, Wallace was the chief spokesman for the Darwinian position. In doing so he had opportunity to explicate to the group some of the difficult points about the relationship of adaptations and natural selection. For example, at the April meeting of 1864, the President, F. P. Pascoe, exhibited some new longicorn beetles sent to him from Australia.[36] He remarked upon their similarity which, he said, might be regarded as "mimetic resemblance" were it not for the fact that the stronger seemed to imitate the weaker. It was his view that it would not be consistant with the view of Bates and Wallace, since a stronger species would not benefit from resembling a weaker species. Wallace explained that advantage did not always depend upon physical strength alone in the struggle for life.[37] A structurally weaker species might have some other means of protection from its enemies which the physically stronger species lacked. In such a case, mimicry of a physically more delicate species by a more robust one might be favored by natural selection.

After the discussion of Pascoe's exhibits, Wallace had some exhibits of his own for the April meeting. He had brought specimens of Malayan Papilionidae to illustrate the phenomenon of mimetic polymorphism. He had spent much time and effort working out the relationships between the male *Papilio polytes* and its various female forms. After his exhibit and explanation, one of the listeners, General John Hearsey, is recorded as having stated that in 1831 he had personally captured *P. pammon* (female) and *P. polytes* (male) *in copula*. He had killed the pair together and had sent them to Professor Westwood at Oxford where they may be seen in that position to this day.[38] That was a bit of data Wallace would unquestionably have liked to have known during his labors to define the relationships between the many forms of *Papilio polytes*.

Before the meeting ended, the subject of local form or permanent variety, as distinct from a species, was again raised.[39] That discussion was continued in the next meeting which featured the participation of the Oxford Professor John Obadiah Westwood and Alfred Russel Wallace. Although it is not possible to determine tone of voice from secretary's minutes, it would seem that the debate was on very friendly terms. Wallace had been explaining the appearance of butterflies taken on Celebes, which made them distinct from relatives on other islands. Wallace had explained that the special Celebesian wing form had been produced through the action of natural selection. The minutes record:

> Professor Westwood, after remarking upon the pleasure he always derived from Mr. Wallace's speculations, whether he agreed with them or not, said that he was unable to follow Mr. Wallace in tracing the phenomenon to the causes assigned by him.[40]

Westwood was not the only entomologist present who doubted the correlation of the curved wing with greater power to dodge in flight. It appears from the number of persons participating that the discussion must have been lively. The participants were mostly not Darwinians, yet their questions were good and to the point. The president, F. P. Pascoe, observed that the theory "Seemed to be based upon the supposition of a gigantic persectuion to which the butterflies had been subject; but was there any evidence of the existence of such persecution?"[41] He asked if Wallace had seen butterflies pursued by birds or other enemies. Pascoe was of the opinion that the level of persecution was not adequate to explain the results attributed to predation. Another anti-Darwinian, Frederick Smith, remarked that "on the persecution principle, great change must be impending in the shape of wing of the butterflies of the London district."[42] Wallace and Bates were almost certainly learning from their opponents. It seems likely that the entomologists who were negatively predisposed to Darwinian evolution were nevertheless very interested in the subject. The descriptive papers did not excite the same kinds of debate. Only the discussion of the priority principle of nomenclature and a few debatable points of criteria for classification of special groups excited sugh lively debates. It would seem as if both the Darwinian advocates and the disparagers of Darwinism were having a good time. Friendliness and not acrimony was the rule rather than the exception. Even J. O. Westwood had an occasional kind word for some of Darwin's work. He did not accept his theory but he knew that the studies he was making were important.

The August meeting in 1865 found a group of both Darwinians and non-Darwinians, including Westwood, gathered around a specimen which had features of both the male and female of its species. The members had been discussing the relationship of the specimen to the different categories of variation that had been suggested. The secretary recorded Westwood's concluding comment as follows:

> With respect to variation generally, no attempt to classify the various forms and phases of it had yet been made; the subject was a wide one, but it would have to be dealt with, and in the hands of a Darwin might be made of surpassing interest and value.[43]

Although one of Darwin's most consistant and articulate critics, Westwood fully recognized Darwin's ability to derive valuable insights from vast quantities of data. It seems likely that Westwood sincerely wished that Darwin would limit his work to the explication of difficult problems like working out the means by which the forms of a polymorphic species are kept from mingling by unknown laws of heredity. Darwin was good at difficult problems and he would be usefully occupied. Westwood recognized that Darwin's excursus into hypotheses was having a wide range of disruptive effects.

By 1865, the question of the origin of species began to be discussed in the more general papers which were reviews of literature in almost any of the entomological groups. One such paper was Robert M'Lachlan's "Observations on some remarkable varieties of <u>Sterrha</u> <u>Sacris</u>, Linn., with general Notes on Variation in Lepidoptera."[44] M'Lachlan had become a reluctant believer in evolutionary divergence and felt compelled to announce this in his concluding general notes. He had become convinced that there were cases in which reversion to type did not occur. He believed there were cases of widening divergence which could divide a group into distinct species. He did not want his hearers to think he had taken up an extreme position. He demurred, "I do not say that I am prepared to accept the 'development theory' to the full extent to which some would apply it; but that it is a reasonable way of accounting for phenomena, which otherwise cannot be satisfactorily settled, must, I think, be evident to all who endeavour to rid themselves of hereditary prejudices." M'Lachlan continued:

> The acceptation, partial or entire, of this theory is not so disastrous as some would appear to consider it. The most inveterate describer of new species need not fear that the darling object of his existence is useless and aimless if well done, for the process of development is of necessity so immeasurably slow, that to all intents and purposes, a faithful description of a new "form" or "species" is as useful to a naturalist holding the one view as to another who prefers to adhere to old ideas; it is only that the one looks upon the origin of that species in a different light from the other.[75]

M'Lachlan was quite correct that taxonomy is as necessary to Darwinian as to a believer in static species. M'Lachlan was, however, playing down the importance many taxonomists placed in their definitions. If evolutionists define "species" and "form" in a new way, it opens up many questions about the sepcies defined according to an older definition. They did not yet know just how disastrous the new theory would be for the taxonomic work already done at great expense, both in effort and money.

M'Lachlan's attempt to be comforting does not seem to have had much effect. The next month's meeting was the Annual General Meeting, at which the President presented his summary of the year's events and their importance for entomology. Francis Pascoe had become aware of the intrusion of Darwinism into many of the sectors of entomological study. He alluded to T. V. Wollaston's brief treatment of the origin of species on islands in his new catalogue of the beetles of the Madeira Islands.[46] It is, however, the importance of the concepts of species and genera that brought him to emphasize the problems which evolutionary theory posed for traditional taxonomy. He was concerned that taxonomy was being undermined by relativistic definitions. It was difficult for Pascoe to say in what philosophical sense genera and species may be said to exist. Pascoe was forced to fall back upon the conventional type concept, although he was keenly aware that it had already suffered serious alterations and taxnonomy was faced with the possibility of its further revision. He wrote uncertainly, "Although genera cannot be sadi to have an absolute existence any more than species and can only be regarded as expressive of an idea, yet as such they have a relative existence."[47] Pascoe was concerned about what became of a taxonomy based only upon designations of convenience, not mandated by nature. Pascoe's presidential address marked the end of the first five-year period following the publication of the Origin of Species. In that period, it had been read, reviewed, and applied to particular studies of entomologists active in the London society.[48]

Later Debates, 1866-1875

Succeeding debate in the years from 1866 to 1875 took place during an extremely active period of research in entomology directly related to evolution. Experimental tests of predator responses to insect prey were devised to stand as tests for natural selection. The decade 1866-1875 also saw three Darwinians in succession elected presidents of the Entomological Society. In this period of Darwinian presidencies, mimicry and other protective devices of animals were often the focus of debates with respect to their relation to the evolutionary theory. The entomologists in the period after 1865 were also confronting difficult philosophical issues. They were having to face the question of the nature of evidence and of proof in respect to theory. It was a period in which some people expressed a real fear of the Darwinian Theory. It seemed to some that "the darling object of his existence was to be rendered useless and aimless"[49] by Darwinism. The taxonomists who had spent much of their working lives contributing to systematic studies of insects believed the new theory boded ill for the future. The emphasis on theory would lead to a devaluation of careful attention to detail. They feared a radical swing of the pendulum too far in the direction of deductive reasoning.

Despite the slowness with which the Society took up the subject of Darwinian evolution, once the topic had been given a place of prominence in both the Proceedings and in the Transactions the references to it became, rather quickly, more and more frequent. The Entomological Society became a good place in which to raise debatable points since debate was guaranteed because Westwood and Pascoe would be there in the company of Wallace and Bates.

Darwin perceived the situation as ideal for his needs. The Entomological Society was perfect for asking specific questions as well as learning what the objections would be to his theory.[50] Darwin could ask questions he needed answered in the writing of his new book The Descent of Man. It was an advantage to be able to direct questions through his friends, Bates and Wallace. Darwin's questions were about insect color and sexual variation which were current topics at the Society any way. The minutes show that the Members were eager to discuss Darwin's questions and even those that were objectors to his theory showed no objection to providing him with their observations.

Three major problems dominated the discussions when the subject of evolution was also discussed. The first was the relationship between the types of variation seen in animals and the permanence or impermanence of species. The second problem was the explication of sexual dimorphism with regard to its origin and its function. The third problem was the relationship of protective coloration, particularly mimetic resemblances, to natural selection.

The bearing of studies of variation on the stability of species was a recurring refrain in many discussions during the 1860's. The degree to which variations revert to type when they were crossed was not well known. Darwin's book <u>Variation of Animals and Plants under Domestication</u> was very timely when it appeared in 1868.[51] Several papers relating to variation and speciation were given in the later 1860's at the entomological societies.[52] One of them "On Species and Varieties," by Captain Thomas Hutton was given as an answer to a Darwinian critic, C. R. Bree.[53] Much of Hutton's paper was not published. The secretary, J. W. Dunning wrote: "The remainder of the paper was a criticism of the Darwinian theory of Natural Selection, the writer's views being principally enforced by arguments beyond the province of the Entomological Society."[54] The parts published dealt with cases in which reversion did not occur. This is the only case in which a paper is denied full publication with the reason being given that it trespassed out of its proper sphere.

The editor of <u>The Entomologist</u>, Edward Newman, was author of another paper on variation. The paper was written in the form of a letter to T. V. Wollaston.[55] Newman perceived that Wollaston was an ally of the anti-Darwinian faction in entomology, though Wollaston had not gone far enough, according to Newman, in rejecting Darwinian views of variation. Newman complained that Wollaston treated variation as "a beautiful and unknown beetle that has just fallen in your path." Newman believed that variation could not be treated as a single, monolithic phenomenon, but ought to be treated as the result of different causes, such as sex or climate. Newman had a characteristic way of writing in metaphor:

> I would rather regard variation as a collection of beetles naturally but accidentally associated, as we sometimes find them after a flood, a sarcophage and phytophage--the lion and the lamb, lying down together, and haply a necrophage--like the vulture or the hyaena--curiously examining whether death had yet adapted them to his peculiar taste.[56]

Newman was concerned that the distinctions between variation with respect to the distinct traits of size, form, color not be obscured any more than the distinction between "normal" and "abnormal" variation like the spontaneous occurrence of albinism in some animals. Newman had not been satisfied by any previous treatment of variation. He found fault with Wollaston's treatment on the grounds that he failed to appreciate the distinctions between forms of variation, but he was even more critical of Darwinian discussions of variation. Newman showed no real understanding of the Darwinian theory. He failed to comprehend many of the distinctions the Darwinians made between variation and adaptation. Newman supposed that cases in which the males of butterflies are larger than the females were contrary to Darwinian expectations ex hypothesi. Newman wanted to warn Wollaston and other entomological scholars away from the Darwinian hypothesis.

> The first difficulty, the primary stumbling-block in our path in every search after truth is hypothesis...On the very threshold of our inquiry into sexual variation we find the Darwinians, or as we may call them the evolutionists, have placed an enormous stumbling-block: they say that the females of insects are larger than the males: in order to establish their views this ought to be so, but it is not.[57]

Newman did not state where Darwinians insisted upon male insects being smaller than female insects. He ascribed a belief to Darwinians that could be easily refuted by observation. Newman apparently hoped that the Darwinian threat could be quickly dispatched by observation. Newman used the term "hypothesis" as a pejorative epithet. Hypothesis united with authority had "delayed the truthful explanation of the phenomena of the solar system," he wrote.[58] The scientific revolution had been sparked by the breaking down of Ptolemaic "hypothesis" with its support in the secular authority of Aristotle and with the sacred authority of the Roman Catholic Church. In Newman's opinion, the destruction of hypothesis had resulted in the triumph of truth in the scientific revolution and it would be another triumph for truth if the Darwinian hypothesis met a similar demise. His hope for a quick dismissal of the Darwinian theory was utterly in vain, especially in the very area he had offered his anti-Darwinian arguments. Sexual variation and dimorphism were among the most frequently discussed topics at the Entomological Society. They were, furthermore, topics in which Darwinian naturalists had much evidence to show in support for natural selection.

Many of the exhibits shown at the meetings were illustrations of sexual differences in insects. Wallace's study of mimicry found only in females had stirred discussion. Mutant butterfly specimens were occasionally found which had male coloration on one or more wings and female coloration on the remaining wings. Darwin's questions had been intended to generate information on the subject of sexual variation for inclusion in his discussion of sexual selection in volume two of the *Descent of Man*.

Newman was not the only entomologist to believe that the matter of sexual variation was a vulnerable point at which to attack the Darwinian hypothesis. That was also the view of John Obadiah Westwood. Westwood brought a drawing of a mutant butterfly specimen from the collection of William C. Hewitson.[59] The butterfly, a *Pieris pyrrha*, had three of its four wings colored in the pattern typical in males and resembling other species of its genus. Its right hind wing, however, was colored like the females of <u>Pieris pyrrha</u>, which meant that it was very different from the other wings, which resembled the Heliconiidae. Professor Westwood is reported to have remarked that such a specimen afforded ground for comment upon Bates's recent paper which had been the subject of much discussion. The unusual butterfly illustrated, Westwood said, how an entomologist can be led into error if he failed to appreciate that even in closely related species, each species is independent of others. He believed that the repetition of colors in the female <u>P. pyrrha</u> of the colors of Heliconiidae was a case of created analogy. Westwood knew that analogous resemblances had been pointed out by W. S. MacLeay and elaborated by W. Swainson.[60] To Westwood, these analogies explained as resulting from the creation of the insect were better explanations than the Darwinian explanation of mimicry.

Westwood explained to the Entomological Society members present what he meant by analogies and gave examples of them. He did not ask about the secondary causes of the analogies, but attributed them to a supernatural cause. Nature had been made to have correspondences and analogies at all levels: for example, between different kingdoms, as is seen in the stick insects corresponding to twigs. Correspondences are also found between classes of the same kingdom in the humming-bird and the humming-bird moth, and between orders of the same class, for instance the drone-fly and the bee. The Darwinians were making their mistake, Westwood believed, in the lower levels of analogies: for example, in their understanding of the correspondences between different families of butterflies, as in the relationship of *Leptalis* butterflies to the Heliconiidae. Westwood furnished no less than seven grounds on which he found the Batesian view of mimicry unreasonable.

To Westwood, it seemed unreasonable to suppose that natural selection could produce mimicry in one sex only. Furthermore, it seemed unlikely that if mimetic resemblances were effective so many unprotected forms thrive without "attempting" to mimic a group like the Heliconiidae. The fact that there were correspondences in color among the allegedly protected species of Heliconiidae opposed the Batesian position. Bates assumed the existence of Heliconiidae prior in time to the arrival of mimic species; that assumption was based on the invalid belief that the species are changing. Westwood saved his strongest argument for last:

> According to the doctrine of chances it was in the highest degree improbable that a casual variation of any given species of *Pieridae* [mimics] should by constant modification, assisted by hereditary descent, gradually assume the form, colour and markings of another species, especially of so remarkable a type as the Heliconiidae. But for an entire

> group to be simultaneously engaged in such a process, each species tending towards distinct and equally peculiar species, would by a logician be pronounced impossible. The admission that the God of Nature created these species in their present condition for some wise but hidden purpose disposes of all difficulty.[61]

For Westwood, an explanation that included God and his Providence was still a better, simpler explanation than the Darwinian hypothesis. If one must resort to explanations based upon unknown forces he would prefer the explanation based upon God's benevolence than "the doctrine of chances."

That same evening Wallace and then Bates responded as best they could to Westwood's objections to mimicry. Wallace laid stress on the advantages conferred on an insect which resembled an object or insect which birds did not attack. He explained why it might be that females of a species might require protection not needed by the males of their species.[62] When Wallace finished speaking, Bates expressed his view that the mutant butterfly on display that evening was irrelevant to the question of the origin of species. The Darwinian theory was not concerned with cases which were clearly not viable in nature. The Darwinian theory was concerned with cases in which variations could be perpetuated by inheritance. He posed a difficult question for Westwood.

> If the dress now worn by the <u>Leptalis</u> was given at its creation how would he [Westwood] explain all these numerous shades of variety found in one and the same locality? To be consistent he must say that each variation was lineally descended from an originally created variety.[63]

When the Darwinians had the floor it was much more likely that the discussion would turn to questions of what happens in the field. Bates tried to get Westwood to think about the diversity of variation observed by a field zoologist and never seen by a museum taxonomist. For Bates, questions ought to be asked about variation in the field where the live organisms are rather than in the museum where the dead specimens are artificially arranged.

The same topic dominated the next meeting, when some new participants joined the debate. On the whole, the discussants showed some good perception of what kinds of data were wanted to support the evolutionary position. The members realized that predation must be on the adult insects if they were to prove mimicry theory and it must be proved that hunters work by visual rather than olfactory cues. On these and other points, the members were willing to offer what they knew of insects from Britain and from other places. Thomas Belt, who was introduced by Bates, gave a "detailed narration" of his observations on the rejection of Heliconia both as adults and as larval stages.[64] The entomologists showed a much poorer understanding of Darwin's <u>Origin of Species</u> and of its application to the phenomenon called "mimicry." One conspicuous example is that of David Sharp, who had prepared his own list of criticisms of Darwinism for the December meeting.[65] A single example of Sharp's faulty comprehension of the nature of natural selection was revealed in his first criticism where he argued that natural selection could not produce resemblances, because it is a "power of differentiation." Natural selection, as Sharp understood it, made species increasingly more different from one another; it could not, then, be the cause of mimetic resemblances in which different species take on similar colors.

Wallace tried to explain to Sharp that natural selection was not only differentiating in its effects; it was a general process of preserving and accumulating useful variations.[66] If it was useful to one species to resemble another, any variation tending toward resemblance could be preserved until enough external resemblance had been accumulated to deceive its predators. The objections raised against evolution by natural selection were answered forthrightly by the Darwinian advocates, Wallace and Bates. But few minds seem to have been changed. The same objections were raised again and again with similar responses. Arguments about theories seem to have had little power to persuade. The addition of experimental data on predation of insects while not <u>proof</u> on natural selection in nature, was the type of evidence the Darwinians needed to bring the discussions back to the kind of inductive basis the entomologists both understood and approved.

The Beginning of Experimentation

Experiments had begun rather inauspiciously with the offering of various kinds of caterpillars to domestic poultry and then observing their responses to them. Insectivores other than birds were also tested for their food preferences. Lizards, frogs and spiders were offered different species of caterpillars.

The very old style natural history had a new Darwinian application. Naturalists, notebooks in hand, recorded precisely the kinds and numbers of prey taken by the residents of their vivariums. An example of the early attempts to verify food preferences in connection with coloration is found in the work of Arthur G. Butler.[67] He recorded exactly what his three green lizards ate from the 3rd to the 13th of July, 1863. They ate 119 large flies, bluebottles, houseflies, and caterpillars, 14 white butterflies, and three humble bees. He went on to observe that by comparison to their usually enthusiastic responses to various foods, the *Lacerta viridis* gave markedly negative reactions to the conspicuously colored insects they were offered. It was an affirmation of Wallace's view that bright colors were useful as warning signals to potential predators. Butler's paper was published in the *Transactions of the Entomological Society for 1869*. He wrote:

> Notwithstanding their eagerness for food of all kinds, from a lemon-cheesecake to a spider,

> there were some caterpillars, and even moths, which they would seize only to drop in disgust... a lizard might spring at the insect when introduced, but after a severe pinch, which did not disable it, but disgusted its captor, it was flung away, and not again disturbed.[68]

Butler was not the only observer engaged in simple experiments John Jenner Weir did similar experiments on birds.[69] The correlation of bright color with rejection by insect predators was clear to the men who kept track of the foods preferred by their captive animals. Jenner Weir's and Butler's papers did not go without objections from entomologists. People enumerated the conspicuous caterpillars they had seen eaten. Nevertheless, Wallace's prediction that bright coloration and palatability were inversely related had been confirmed in a test. In most caterpillars, bright color has been preserved as a warning color signalling unpalatability.[70]

John Jenner Weir's series of experiments had been suggested to him in April 1867 by Wallace, and he was glad to acknowledge in the discussion at the Entomological Society that his results confirmed Wallace's prediction.[71] Wallace was especially pleased that conspicuous color in insects could be explained by the action of natural selection as distinct from sexual selection.

It is not surprising that Bates made particular note of the short papers by Butler and Weir in his Presidential address for 1869.

> The volume for the past year comprises twenty-seven memoirs, of which twenty-five belong to

> the department of systematic or descriptive Entomology, and two only--welcome contributions from Mr. Jenner Weir and Mr. Butler, on the selection of insects as food by insectivorous animals--to other branches of the Science.[72]

Bates expressed hope that even writers of descriptive papers would make it a more and more frequent practice to include data on the relationships between faunas, on geographical distribution of species, and "other deeply interesting philosophical questions."[73] Bates's tenure as president of the society was followed by two years of Wallace's presence in the chair. Darwin's controversial book on _The Descent of Man_ was published during Wallace's term as president. The subjects Darwin explored, especially sexual selection, sexual coloration and the evolution of unusual reproductive strategies, were vigorously discussed during Wallace's presidency in 1870 and 1871.[74] Also, the subject of mimicry continued to be debated. More interest was shown in topics dealing with geographical distributions and their relationship to evolutionary theory throughout the period of Darwinian occupancy of the president's chair at the Entomological Society.

Peaceful Coexistence between Darwinians and Anti-Darwinians

There was no _coup d'etat_ at the Entomological Society. The Darwinians did not seize the positions of importance or revolutionize the institution of the Entomological Society of London. Darwinians and anti-Darwinians served together on the Council and as Vice-Presidents. Just as F. P. Pascoe was succeded by John Lubbock as president, Wallace was succeded by an anti-Darwinian, J. O. Westwood. Westwood had been a member of the Council during the presidencies of both Lubbock and Bates; Bates was a council member while Westwood was president. A non-Darwinian might serve as chairman at a meeting due to the absence of a Darwinian president and vice-versa. There was no apparent polarization of the society or unwillingness on the part of either Darwinian advocates or opponents to serving together in making decisions on the behalf of the Entomological Society. It was still true, perhaps naturally, that the liveliest discussions of evolution were held when a Darwinian was in the chair. Theperiod of Darwinian presidents coincides with the most active debate of the evolutionary theory in the scientific community at large. The discussions at the Entomological Society, preserved in the Society's <u>Proceedings</u>, suggest that the Darwinian theory was given an open forum in the meeting, perhaps the most open of any of the scientific societies in London. The papers published in the <u>Transactions</u> dealt with topics of obvious bearing on the issue of evolution in a more formalized and often in a euphemistic manner.

It is also important that many of the members of the

Entomological Society who published descriptive papers in the Society's Proceedings had papers on evolutionary topics published in the Transactions of the Linnean Society. It is not entirely clear that they made that choice because of an uncongenial atmosphere at the Entomological Society or a hostile editorial policy. It seems more probable that the authors of papers on evolutionary subjects were seeking the wider circulation and the greater prestige associated with the Transactions of the Linnean Society. Bates's paper on mimicry was published there as was Wallace's and, later, Roland Trimen's. Edward Poulton was probably correct in suggesting that the determining reason may have been the quarto size of the volumes and the advantages that gave in the publication of plates.[75]

It may have been that the Entomological Society seemed like the place to offer technical, descriptive works. Many of the most prominent members were descriptive taxonomists who appreciated studies of particular groups. Upon reading the descriptive papers, one frequently finds that they contain matters of "philosophical interest" as Bates would have put it.

Many of the authors of descriptive papers were well aware that their work had implications for or against the evolutionary theory. It was still rather rare to reveal that relevance in the titles. Tables of contents are very misleading guides to the history of the reception of the evolutionary theory.[76] No papers ought to be expected with a title like, "The Application of the Darwinian Hypothesis to Entomology," or "Variation in the Genus *Limenitis* and its Relationship to the Theory of Mr. Darwin." The papers bear the plainest headings and yet contain very provocative material for the Darwinian debate. Even the Darwinians chose descriptive titles which hid rather than revealed the theories which were stated, or at least were implied within the paper. Bates, who was a master at selecting emotionally neutral titles for his papers, used the general title of "Contributions to an Insect Fauna of the Amazon Valley" for several important evolutionary papers. For smaller papers which discuss evolution he used the designation "Notes": for example, "Notes upon the Variation of Sexes in *Argynnis Diana*."[77]

It seems likely that the Darwinians were behaving according to an unstated code of propriety. They undoubtedly also perceived it as in their own interests to show the inductive entomologists that they were not only concerned with theory. They were also committed to the gathering of facts and equally dedicated to the presentation of the facts taken from Nature. Their opponents used the same facade of theoretical neutrality. Under a plain title a body of data was given and then, and only then, in the center, or at the end of a paper, a point or two was made which challenged the evolutionary position. The opposition was not likely to dignify the Darwinian theory by naming it in the title. The standard organization for papers was to put the data first, and the theoretical implications to be taken from them, then appear to emerge naturally--inductively--from the amassed data.

One example to illustrate the style of arguing against general evolutionary conclusions is found in a paper devoted mainly to descriptive taxonomy, by John Edward Gray, a past president of the Entomological Society. He opposed Darwinian claims in a paper of 1863 entitled "Revision of the Species of Lemuroid Animals, with the Description of Some New Species."[78] In that paper, he made his position perfectly clear. If species derive gradually from ancestral species, he had no reason to think so. Furthermore, he had some forty years of experience with specimens which was perhaps more than anyone else and he should know. John Gray was not a man to equivocate. The following year he had another paper published called "Revision of the Species of Trionychidae Found in Asia and Africa with the Descriptions of Some New Species." He found space in his paper on a group of tortoises to strike a blow against Bates's theory of mimicry in butterflies.[79]

In the first fifteen years of discussion of Darwinian evolution, it might be said that zoologists did not object to discussing the relationship of Darwinism to their particular data, but they did not want to appear to be discussing theory except in the context of a body of data.

The restraint seen in the published papers was relaxed in the society meetings both in the period of the Darwinian presidencies and in the period that followed Wallace's retirement from the presidency in January of 1872, when J. O. Westwood was elected president for the second time. Westwood was a man of great tolerance who did not ridicule the positions he opposed. He was committed to a position of permanent, created species, and could not be persuaded to change his position. However, he was beginning to lose some of his patience by 1873. It was apparent to him that the teaching associated with the Darwinian theory of evolution was a subject that must be squarely combatted. For Westwood, Darwinism was flawed both scientifically and morally. In his Presidential Address for 1873, he made the Fellows of the Entomological Society aware of a letter he had recently written to the Vice-Chancellor of Oxford University.[80] His letter appears in its totality in the published address and concludes with the following statement.

> I cannot conclude this letter without mentioning another subject which has long forced itself on my attention, namely the want of a Professor or Reader in Natural Theology, whose duty it should be to counteract the atheistical demoralization resulting from the unlimited teaching of Darwinism, in which Design in Creation, and even Creation itself, are openly or virtually denied or ignored.[81]

Westwood was a consistent disbeliever in the evolutionary theory; he considered it a dogma which carried with it an unworthy picture of the Creator or a denial of His very existence.

Despite his differences with the Darwinians he retained their high regard. He was elected by acclamation life-president of the Entomological Society at the Jubilee Meeting which marked the Fiftieth Anniversary of the Society.[82] In its first century the Entomological Society named its two most famous advocates of Natural Theology its only two titular Life Presidents.

The election of Professor Westwood to the Presidency for the years 1872 and 1873 was followed by the election of Sidney Smith Saunders for 1874 and 1875. Westwood was elected again in 1876 and 1877. The six year period of Darwinian chairmanship was followed by an equally long period of anti-Darwinian chairmanship of meetings. It may not have been their anti-Darwinian position which was the reason for their election. Both Westwood and Saunders were original members and it was an appropriate gesture for the society to make on the basis of their long and faithful interest in both entomological studies and in the furthering of the society. It probably was not pleasant, however, for the Darwinians to hear their position characterized year after year as an untenable assumption. Sidney Smith Saunders was more scathing in his appraisals of evolutionary theory than Westwood had been.

Saunders had the last word on the first fifteen years of debate on Darwinism. He said:

> It has been the fashion of the day to elaborate new theories of development from antecedent epochs, and to attribute to adventitious circumstances all the marvellous results we now witness throughout Nature's realm. When, however, it becomes essential to reconcile hypothetical arguments with the stern logic of facts, we find the Coryphei of the Evolutionary Propaganda at variance among themselves on fundamental principles.[82]

Saunders believed the inner dissension among Darwinians indicated an irreparable flaw in the theory. Saunders preferred to deal with the Darwinian question as a study in social rather than intellectual history. Disagreements among people were used to avoid confronting their ideas.

In some ways the end of Darwinism's first fifteen years was not unlike its beginning. Sidney Smith Saunders's vituperations from the chair are not unlike those of the president of 1859, John Edward Gray, although Gray saved his criticisms until he was out of the chair. Darwin was quite correct in his opinion that his enemies would have to die and be replaced by younger people who could understand a new theory by which to guide their researches. The acceptance of his theory really did "take two or three lifetimes" as he had predicted it would. He was equally correct when he wrote, "The entomologists are enough to keep the subject back for half a century."[84]

The full acceptance of Darwin's theory did not occur during Darwin's lifetime even at the Entomological Society where he had such able advocates. Henry Walter Bates lived long enough to begin to see important changes in entomological studies that acknowledged evolution of new species from pre-existing species. It was really only Alfred Russel Wallace, who lived until 1913, who lived to see evolutionary theory acknowledged by researchers as their major research principle. However, these were still very difficult times for those whose belief in general transmutation of species was accompanied by a strong position on natural selection as its primary agency. At the close of the nineteenth century the almost exclusive emphasis on taxonomy in entomological circles was beginning to yield to more diversity in research areas.

The subject of insect coloration was no longer only studied after the traditional manner of natural history. The protective coloration of animals was being studied under experimental conditions by Edward Poulton.[85] The pigments of color were being analyzed by several researchers[86] and August Weismann was seeking the basis for evolution of markings through meticulous morphological studies.[87]

Raphael Meldola commented upon the manner in which entomological studies had changed when he delivered his

Presidential Address in 1895.[88] In his lecture he applauded the modernization of entomology, which for him meant the adoption of methods of thinking theoretically and of using improved theories in the devising of experiments. He believed that the old "fact-gathering" natural history approach had been an infantile stage of development which entomology ought rightfully to outgrow. Advances in recent physics demonstrated the advantages of taking up their new theoretical approach. Meldola quoted a physicist who said, "It is the rejected theories which have been the steps toward formulating others nearer the truth."[89] Meldola wanted to encourage entomologists to cultivate their idle imaginations. In the modern world speculation would be encouraged rather than denigrated. Meldola recalled that Bates first speculated on mimicry in butterflies after he had departed from the Amazon, "while looking over his specimens when he reached London, and was pondering, at his own fireside, over the memory of the remarkable superficial resemblances among the butterflies of differemt groups which he had brought home."[90] Bates's speculations had had the good result of catalyzing further observation. Müller's observations and theoretical conclusions had been based on modifications of Bates's theories, and then in turn, C. Lloyd Morgan's studies of animal behavior were stimulated by Müller's hypotheses.[91] Meldola continued:

> As the product of a theory we thus have a large body of real and tangible knowledge gleaned from Nature! Mere casual observation would never have revealed the widespread existence of the phenomena if stimulus to look for it had not come from the theoretical side.[92]

Theories about insect coloration exemplified the salutary effect which good ideas have when they are skillfully applied to experiments.

Wallace's theory of warning coloration was one such good idea which had been subjected to experimentation and had been experimentally verified. At the beginning of his work on warning colors, Wallace had been working out the various uses for color and trying to define their relationship to natural selection as distinct from Darwin's work on colors which emphasized sexual selection. The experiments of J. Jenner Weir and A. G. Butler in the late 1860's confirmed the relations between consqicuous color and unpalatability to predators. It was these kinds of fruitful interactions between theory and experiment that appealed to Meldola. He was impressed with the way in which ideas pass from mind to min, increasing in their value through imaginative theories and experimental verifications. He remembered that Bates's theory of mimicry had inspired Thomas Belt to collect data on insect markings which, in turn, were developed by August Weismann into an important work on the biological meaning and origins of markings in butterfly larvae. Following

Weismann, E. B. Poulton's studies were enriched by a legacy of ideas which went back through a chain of theorizers to Wallace and to Bates.

> The colour variability of caterpillars and pupae in response to the external stimulus exerted by coloured surfaces, as established by these experiments, had brought us face to face with a fundamental problem in insect physiology, the solution of which we are anxiously awaiting. The mere possibility of being able to state the problem in its present form--apart from any questions of the adaptive value of the colouration--is a step forwards; is an incentive to further experiment and this is the legitimate end and aim of all scientific speculation.[93]

Meldola believed that if biology would have the power and prestige of physics, it must exploit its methods. With a clever turn of phrase, he stated, "It is better to have speculated erroneously than never to have speculated at all."[94]

Meldola, who represents the "biologist" as distinctly evolved from an ancestral "naturalist," was not speaking for all entomologists. There were those who recalled that Bates refused to generalize even on the better known groups of beetles in his last studies of Coleoptera in the Biologia Centrali-Americana.[95]

The fifteen years following the publication of the Origin of Species was a period in which the rift between taxonomy and the new biological sciences began to be a significant factor in the discussions at the Entomological Society. The Society was a liberal institution which permitted almost any opinion to be aired. The price of that freedom was often division. Darwin's opinion that the world is divided into two kinds of minds--those that are inductive and those that are deductive--is attractive as a model although it is too simple. The proponents of induction were right that much work still remained even to merely identify the vast assortment of "organized beings" which were still unnamed. It was the opinion of most inductive taxonomists that only after an indefinitely long period of fact-gathering could meaningful generalizations be made about the facts. Many believed further that the scientist had a moral as well as a scientific duty to describe Nature as it is. As spokesmen for a more deductive methodology, the Darwinians had discovered the power of a good theory for organizing bodies of data. The logical force of deduction, to the Darwinian, seemed far stronger than that found in induction. Their importation of more rigorous experimental methods into biology in conjunction with Darwinian theory was welcomed by many younger entomologists. Nevertheless, the advocates of the older style of natural history could agree with the new biologists on a summation of the state of science in England which was written by an anonymous writer in the "Standard": "In this country biological science is in a state of transition. The old style is passing away; the new has not yet fully developed."[96]

Notes--Chapter IV
EVOLUTION AT THE
ENTOMOLOGICAL SOCIETY OF LONDON, 1860-1875

1. The list of original members is found on pages 162-164 of *The Centenary History of the Entomological Society of London*, 1833-1933, by S. Neave. (London: Richard Clay & Sons, 1933).

2. A list of ordinary members and subscribers was printed in the *Transactions* of the Society for 1858 as of October of that year, pp. xix-xxiv. Another list appeared as of October, 1861, in the *Transactions* for 1861, pp. xvii-xxi.

3. The prevalence of clerical and military titles is one type of clue to their amateur status. The little obituaries given of members is a better way to judge whether a person was an amateur or not. *The Entomologist* and the *Entomologist's Monthly Magazine* were more consistent in giving biographical details than were the Presidents of the Entomological Society in their "memorial minutes."

4. Henry Tibbats Stainton (1822-1892), "How may the Onward Progress of the Study of Entomology be best furthered?" *Transactions of the Entomological Society of London* 4 (2): 138-142, 1856-1858. (Hereinafter abbreviated *Trans. Ent. Soc.*).

5. Stainton, "How...furthered?" p. 371.

6. Letter, Charles Darwin to Hugh Strickland, February 4, 1849. *Life and Letters of Charles Darwin* I, pp. 369-371. (Hereinafter referred to as *Life and Letters*).

7. Letter, Darwin to Strickland, February 4, 1849. *Life and Letters*, p. 371.

8. Letter, Darwin to J. D. Hooker, April 9, 1849. *Life and Letters*, pp. 375-376.

9. Letter, Darwin to Hooker, September 25, 1853. *Life and Letters*, II, pp. 39-41.

10. Letter, Darwin to Hooker, September 25, 1853. *Life and Letters* II, p. 40.

11. Letters from travelers appeared both in the *Proceedings of the Entomological Society* and in *The*

Zoologist. The same letter, for example, appeared in the *Proceedings* for 1856 on pp. 10-11 and pp. 91-93 and on pp. 5657-5663 of *The Zoologist* for 1857 and pp. 6160-6169 for 1858.

12. A. R. Wallace, "Note on the Theory of Permanent and Geographical Varieties," *Zoologist* *16*: 5887-5888, 1858.

13. Wallace, "Note...Geographical Varieties," p. 5888.

14. Bates, "Notes on South American Butterflies," *Trans. Ent. Soc.* *5*: 1-11, 1858-61. Read September 7, 1857.

15. Bates, "Notes on South American Butterflies," pp. 1-2.

16. Henry Walter Bates, "Contributions to an Insect Fauna of the Amazon Valley. Part I. Diurnal Lepidoptera," *Trans. Ent. Soc.*, 1858-61, pp. 223-228 and pp. 335-361. The second part of the paper, read November 24, 1860, contains the reference to the *Origin of Species*, Chapter 9, p. 378, on p. 352.

17. John William Douglas (1814-1905), [Presidential Address], *Proc. Ent. Soc*, 1862, p. 38. Address delivered January 28, 1862.

18. Douglas, [Presidential Address], p. 38.

19. Darwin's first edition, *On the Origin of Species*, pp. 109-11, 126, 317-333, 431-433.

20. The Anniversary Address for 1859, given January, 1860, is in the *Trans. Ent. Soc.* *5*: 91-100, 1858-61.

21. William Forsell Kirby (1844-1912), "On the Geographical Distribution of European Rhopalocera," *Proc. Ent. Soc.*, 1862-64, pp. 481-492. Read January 5, 1863.

22. William Forsell Kirby, *Manual of European Butterflies* (London: Williams and Norgate, 1862).

23. W. F. Kirby, "On the Geographical...Rhopalocera," p. 482.

24. W. F. Kirby, "On the Geographical...Rhopalocera," p. 482.

25. W. F. Kirby, "On the Geographical...Rhopalocera," p. 491.

26. Proc. Ent. Soc., January 5, 1863, p. 115.

27. An example of a reference to mimicry in this period is that made by T. W. Wood on May 4, 1863. Proc. Ent. Soc., p. 147.

28. Frederick Smith (1805-1879), [Presidential Address], Proc. Ent. Soc, 1863, p. 131. Address given January 26, 1863.

29. Frederick Smith, [Presidential Address], Proc. Ent. Soc. 1864, pp. 209-215. Address given January, 1864.

30. Frederick Smith, "On the Construction of Hexagonal Cells by Bees and Wasps," [Read April 4, 1864], Trans. Ent. Soc., 1864, pp. 131-142.

31. Rev. Samuel Haughton, "On the Form of Cells Made by Various Wasps and by the Honey Bee; with an Appendix on the Origin of Species," Annals and Magazine of Natural History 11 (3): 415-429, June, 1863. Alfred Russel Wallace, "Remarks on the Rev. S. Haughton's Paper on the Bees' Cell, and on the Origin of Species, Annals and Magazine of Natural History 12 (3): 303-308, October, 1863.

32. F. Smith, "On the Construction...," pp. 132-133.

33. Proc. Ent. Soc., April 4, 1864, pp. 18-20. Edward Newman's (1801-1876) reference to the Microgaster genus as exemplifying the intent of the divine Architect is interesting. The Microgaster is an ichneumon wasp whose "larval cells" are the bodies of other insects which are being injected with eggs which develop and parasitize a host and then kill it. The problem of parasitism was a very serious one for those who argued from benevolent Providential design. See p. 19.

34. Proc. Ent. Soc. 2: 29, June 6, 1864.

35. Proc. Ent. Soc. 2: 104-105, July 3, 1865.

36. Proc. Ent. Soc. 2: 14, April 4, 1864.

37. Proc. Ent. Soc. 2: 15-16, April 4, 1864.

38. Proc. Ent. Soc. 2: 16, April 4, 1864. For the forms of Papilio polytes see Plate III.

39. Proc. Ent. Soc. 2: 16, April 4, 1864.

40. *Proc. Ent. Soc.* 2: 22, May 2, 1864.

41. *Proc. Ent. Soc.* 2: 22, May 2, 1864.

42. *Proc. Ent. Soc.* 2: 22, May 2, 1864. Wallace remarked that the wing of a butterfly was more liable to variation than the body of the insect. The wing was, therefore, more susceptible to changes through selection. It is in this meeting that Wallace admitted that he had not seen butterflies pursued "to any great extent" and that he based his suggestion on theory, but that he had no doubt that predation occurred. His opponents remembered his statement.

43. *Proc. Ent. Soc.*, 1865, p. 110, August, 1864. There is the possibility that Westwood is being ungenerous to Wallace or Bates. It seems more likely that Westwood was expressing the wish that Darwin would write a work on variation in nature in the same vein as his work on domestic variation.

44. Robert M'Lachlan (also sometimes seen as McLachlan, 1807-1904), "Observations on Some Remarkable Varieties of *Sterrha Sacria*, Linn., with General Notes on Variation in Lepidoptera," *Trans. Ent. Soc.*, 1865, pp. 466-467, December 4, 1865.

45. M'Lachlan, "Observations on Some Remarkable...," pp. 466-467.

46. Thomas Vernon Wollaston (1822-1878) had just published his book, *Coleoptera Atlantidum, Being an Enumeration of the Coleopterous Insects of the Madeiras, Salvages, and Canaries* (London: John Van Voorst, 1865).

47. Francis Polkinghorne Pascoe (1813-1893), [Presidential Address], *Proc. Ent. Soc.* 2: 146, January, 1866. Pascoe was struggling with the growing difficulty of how to think about taxonomic levels when there was declining confidence in their reality except as conventions. Genera level decisions are becoming difficult without rigid sets of technical criteria.

48. *The Origin of Species* was reviewed by three entomologists: John O. Westwood, "Mr. Darwin's Theory of Development," in the *Gardener's Chronicle* for February 11, 1860 and reprinted in the *Annals and Magazine of Natural History* 5: 347-348, 1860; Thomas Vernon Wollaston, "On the Origin of Species by means of Natural Selection; or, the Preservation of Favoured Races in the Struggle for Life," [Bibliographical Notice], *Annals and Magazine of Natural*

<u>History</u> <u>5</u>: 132-143, 1860; and Andrew Murray (1812-1878), "On Mr. Darwin's Theory of the Origin of Species," <u>Proceedings of the Royal Society of Edinburgh</u> <u>4</u>: 274-291, 1857-62.

49. McLachlan, "Observations...", p. 466.

50. Questions which Darwin asked were put forward to the group at least three times: March 4, 1867; February 17, 1868; and March 2, 1868.

51. Charles Darwin, <u>Variation of Animals and Plants Under Domestication</u>, 2 vols. (London: John Murray, 1868).

52. Captain Thomas Hutton, "On Species and Varieties, <u>Proc</u>. <u>Ent</u>. <u>Soc</u>., 1867, pp. lxxxii-lxxxiii, March 18, 1867.

53. Dr. Charles Robert Bree (1811-1886), author of <u>Species not Transmutable, nor the Result of Secondary Causes</u> (London: Groombridge and Sons, 1860) and <u>An Exposition of Fallacies in the Hypotheses of Mr. Darwin</u> (London: Longman Green, 1872).

54. Joseph William Dunning (1833-1897). The argumentative nature of a paper would have justified its being published in the <u>Proceedings</u> rather than in the <u>Transactions</u>. The remark is peculiar because criticisms were not suppressed either before or after that day, March 18, 1867. No limits of a formal kind were drawn up to set the proper "province" of the Entomological Society.

55. Edward Newman, "Letters on Variation in Lepidoptera. Letter the First.--Introduction.--Sexual Variation," <u>Zoologist</u> <u>2</u>: 721-727, 1867.

56. Newman, "Letters on Variation...," pp. 721-722.

57. Newman, "Letters on Variation...," p. 724.

58. Newman, "Letters on Variation...," p. 724.

59. <u>Proc</u>. <u>Ent</u>. <u>Soc</u>. <u>5</u>: xxxvi-xli, 1865-67. November 19, 1866. Sir John Lubbock was in the chair.

60. William Sharp MacLeay (1792-1865) was associated with a classification system called the Quinary System. The system purported to be based on analogies built into the animal kingdom. His system was developed for insects in <u>Horae Entomologicae</u> (London, 1819). William Swainson applied the Quinarian system to birds.

61. Proc. Ent. Soc., November 19, 1866, p. xxxviii.

62. Proc. Ent. Soc., November 19, 1866, pp. xxxviii-xxxix.

63. Proc. Ent. Soc., November 19, 1866, p. xl.

64. Thomas Belt (1832-1878) was author of The Naturalist in Nicaragua (London: John Murray, 1874). Proc. Ent. Soc., December 3, 1866, p. xlv.

65. Proc. Ent. Soc., December 3, 1866, pp. xlv-xlvi.

66. Proc. Ent. Soc., December 3, 1866, pp. xlvi-xlvii.

67. Arthur Gardiner Butler (1844-1925), "Remarks upon Certain Caterpillars &c., which are Unpalatable to their Enemies," Trans. Ent. Soc., 1869, pp. 27-29. Butler studied the green lizards (Lacerta viridis) in 1864 and 1865 and frogs' responses to caterpillars in 1868.

68. Butler, "Remarks upon certain caterpillars...," p. 27.

69. John Jenner Weir (1822-1894), "On Insects and Insectivorous Birds; and Especially on the Relation Between the Colour and the Edibility of Lepidoptera and their Larvae," Trans. Ent. Soc, 1870, pp. 227-339.

70. Proc. Ent. Soc., March 1, 1869, p. vii. No claim was made that brightly colored caterpillars were absolutely exempted from predation, but only that they were relatively more free than cryptically colored insects. No really significant statistical studies were then available on the degree of predation associated with different degrees of conspicuous color. Jenner Weir's and A. G. Butler's experiments were exceedingly informal . They were testing the theory Wallace had explained to them only in a very simple, non-biometrical way.

71. J. J. Weir, "On Insects and Insectivorous Birds," p. 21, and Proc. Ent. Soc., March 1, 1869, p. vi.

72. H. W. Bates, [President's Address] Proc. Ent. Soc., 1869, pp. xxxiii-xlix. Quotation is on page xxxiv.

73. Bates, [President's Address] 1869, p. xxxiv.

74. One very interesting paper published by the Entomological Society is B . T. Lowne's "Observations on

Immature Sexuality and Alternate Generation in Insects," read March 6, 1871. (Trans. Ent. Soc., 1871, pp. 193-202). It is a study of the relationship between neoteny in insects and nutritional status. Lowne had been in contact with Darwin and refers to the Descent of Man--one of the earlier references to this work in a scientific paper.

75. Edward Bagnall Poulton (1856-1943), "The Influence of Darwin upon Entomology," Entomologist's Record and Journal of Variation, 1901, pp. 72-76. Quotation from pp. 74-75.

76. See the chapter by Frederick Burkhardt, "England and Scotland: The Learned Societies," in Comparative Reception of Darwinism, edited by Thomas Glick (Austin: University of Texas Press, 1974), pp. 32-74.

77. Henry Walter Bates, "Notes upon the Variation of Sexes in Argynnis Diana," Proc. Ent. Soc. of Phil. 4: 204-207, 1865. Bates explicated to American readers the relationship between sexual and natural selection. He named specific areas in which research needed to be done.

78. John Edward Gray, "Revision of the Species of Lemuroid Animals, with the Description of Some New Species," Proc. Zoological Soc., 1863, pp. 129-152. See especially pp. 134-135.

79. John Edward Gray, "Revision of the Species of Trionychidae Found in Asia and Africa, with the Descriptions of Some New Species," Proc. Zoological Soc., 1864, pp. 76-98. See p. 77. Gray wrote: "I believe that such cases [of accidental resemblance] are much more common than has hitherto been suspected; and it is on such superficial resemblances that Mr. Bates's observations and theories respecting the Brazilian Butterflies are founded--notions which will vanish into the air when the insects are more carefully examined by a systematic entomologist." Gray was very unpleasant to Bates, probably mainly because he was not, by Gray's definition, a professional. Gray also berated Wallace for daring to criticize Rev. Samuel Haughton who was older and better educated than he. See Wallace, My Life, Vol. 2, p. 88.

80. John Obadiah Westwood, [Presidential Address] Proc. Ent. Soc., January 26, 1874, pp. xxv-xxvi. Letter was dated November 27, 1873. It was written to request several new posts for the teaching of Zoology at Oxford. Westwood requested that an Assistant Professor of Zoology, and Assistant Professor of Comparative Anatomy and a Readership in Paleontology be created, in addition to the position in Natural Theology.

81. Westwood, [Presidential Address] 1873, p. xxvi.

82. Proc. Ent. Soc., 1883, pp. vii-xi. J. W. Dunning in the chair, May 2, 1883.

83. Sidney Smith Saunders (1809-1884), [President's Address] Proc. Ent. Soc., 1875, pp. xxx-xli, and January, 1876, p. xxxv.

84. Letter, Darwin to Charles Lyell, March 17, 1863. Life and Letters III, p. 17.

85. Edward B. Poulton, "The Experimental Proof of the Protective Value of Colour and Markings in Insects in Reference to Their Vertebrate Enemies," Proc. Zoological Soc., 1887, pp. 191-274.

86. Several people were beginning to work in pigment chemistry of invertebrate animals. Among them were F. H. Perry Coste, F. Gowland Hopkins, E. B. Poulton, C. A. MacMunn, M. I. Newbegin.

87. August Weismann, Studies in the Theory of Descent, translated by Raphael Meldola (London: Sampson Low, Marston, Searle & Rivington, 1882). Part II. The Final Causes of Transformation. The origin of the markings of caterpillars, pp. 161-554.

88. Raphael Meldola, [Presidential Address], Proc. Ent. Soc., 1895, pp. lii-lxviii. Given January, 1896

89. Physicist quoted was W. M. Hicks, President of Section A of the British Association. Quotation was taken from his address to the BAAS, Ipswich, 1895, p. liv.

90. Meldola, [Presidential Address] 1895, p. lvi.

91. Meldola, [Presidential Address] 1895, p. lvii. One of C. Lloyd Morgan's books which discusses mimicry is Animal Life and Intelligence (London: Edward Arnold, 1890-91).

92. Meldola, [Presidential Address] 1895, p. lvii.

93. Meldola, [Presidential Address] 1895, p. lix.

94. Meldola, [Presidential Address] 1895, p. lxiii.

95. Bates wrote the sections of Coleoptera in Biologia Centrali Americana by Frederick Godman and Osbert Salvin. H. W. Bates, Biologia Centrali Americana. Insecta:

Coleoptera (London: R. H. Porter, 1881-84).

96. Quoted by Robert M'Lachlan in his Presidential Address for 1885. *Proc. Ent. Soc.*, p. xlvii.

Chapter V

ANOTHER DAY'S LIFE: DARWIN AND WALLACE ON SELECTION

Differences of opinion between Darwin and Wallace have been scrutinized by biological workers who are friendly to the evolutionary theory as well as by those who rejoice when the Darwinian house is divided. Some of the liveliest interest in the points of disagreement between Darwin and Wallace is found in the writings of naturalists who are themselves ambivalent about the role of natural selection in evolution.

One noteworthy example of a naturalist who could not believe that natural selection alone was adequate to effect evolutionary change is George John Romanes (1848-1894).[1] For him, the *true* Darwinian was the evolutionist who accepted the same supplementary causes of evolutionary change Darwin accepted. Romanes denied that the title "pure" or "neo-" Darwinian properly described naturalists who held that natural selection was the only means of evolutionary change. The name Romanes used to designate selection advocates was "Wallaceists."[2] In Romanes's opinion, "Wallaceists" were not purer than Darwinians, but only narrower.

The keynote in Romanes's exposition of the differences between the two founders of evolutionary theory was that

Darwin was the more open to the consideration of diverse means and secondary causes producing change. Romanes was particularly whole-hearted in his support of Darwin's view of sexual selection in the second volume of his three-volume work, Darwin and After Darwin, of 1895.[3]

The term "Wallaceism" was being used in an almost accusational sense by the fiftieth anniversary of the Linnean Society Papers in 1908. A. A. W. Hubrecht (1853-1915) wrote an article entitled "Darwinism *versus* Wallaceism" for the 1908 Contemporary Review.[4] Hubrecht, like Romanes, found Darwin's rather eclectic evolutionary theory more to his liking. Hubrecht was eager to demonstrate that he and Hugo DeVries were more staunch in their devotion to Darwin's teachings than either Wallace or his colleagues William Thiselton Dyer and Edward Poulton.[5] Hubrecht touted DeVries's studies of the evening primrose, *Oenothera Lamarckiana*, as true Darwinism. He expounded:

> For those who have really read, perused and digested deVries' wonderful book, as also his earlier "Intracelluläre Pangenesis," there can be no doubt of this; and now that the semi-centennial celebration of the publication of the "Origin of Species" is approaching, it should, in justice, be loudly proclaimed that deVries, an early Darwinian of the purest water, is closer to the original fountain from which Darwinism has sprung than is even Darwin's associate, Dr. Wallace, who has drifted far away in his personal interpretations of variation.[6]

Darwin's name was invoked by proponents of *mutations-theorie* as encompassing all forms of variation, inheritance, and selection.

Of course, Wallace replied to Professor Hubrecht's allegations in the next issue of *Contemporary Review*.[7] Wallace compiled a series of quotations from Darwin's *Variation of Animals and Plants under Domestication* and from the fourth edition of the *Origin of Species* to refute Hubrecht's claim that any antagonistic views were ever held separating him from Darwin on the role of large "single variations" in evolution. Neither had ever held that mutations, large jumps, had more than the most exceedingly minor roles in evolution. Despite his eighty-five years in age, Wallace had retained his acument for evolutionary debate. Whenever there was a hint of disunity between Darwin and Wallace, proponents of alternative evolutionary and philosophical systems were quick to notice.

Darwin-Wallace Letters

The very real opposition of views on the evolution of human beings naturally provoked interest among those who favored evolution and among those who opposed it. Wallace's growing doubts about the adequacy of natural selection to explain human origins were bruited as indicating the failure of naturalistic explanation even for evolution's chief advocates. It was only with regret that Wallace and Darwin maintained varying views, particularly in public. However, as a matter of principle, Wallace believed that he should not conceal his opinions that extra-ordinary causes had been operating in human evolution.[8] Wallace's papers on man were moral statements for him, just as Darwin's Descent of Man was written when he felt that he ought not to withhold his opinions from publication on such an important subject. Even so, fewer pages of correspondence between Darwin and Wallace were occupied with the subject of human evolution than with another subject--animal coloration. Perhaps each held some lingering hopes of being able to persuade the other of the force of his particular theory of animal color, but on man their different views were too fundamentally opposed.

The data each researcher chose to provide the examples of adaptive and sexual coloring was responsible, in large part, for the differences between them in the theories of the evolution of colors. Wallace developed his views mostly from entomological researches. His ornithological observations of wild birds supplemented his work on Malayan lepidoptera. Darwin's sources were different; he had devoted two full volumes to studies of variation in domestic animals and cultivated plants. His work on domestic pigeons and poultry provided him with the basic understanding of the importance of sexual behavior in the production of sexual dimorphism and the legitimacy of the analogy of domestic animals to understand wild species.[9] Four chapters in the Descent of Man were devoted to a very full description and analysis of sexual coloring in birds, wild and domestic.[10] Darwin's understanding of sexual dimorphism in butterflies and other insects was in major part a direct extension of observations and theories developed from work with birds. It is not surprising, therefore, that Darwin's opinions were quite different from Wallace's view of insect coloring based on mimicry and polymorphism studies of butterflies.

Coloration, sexual and adaptive, figures prominently from the beginning of Darwin's and Wallace's publication on natural selection. Darwin's first definition of sexual selection is to be found in their joint Linnean Society publication of 1858. Darwin had declared that secondary sexual characters of animals were the results of an auxilliary mechanism in addition to natural selection and distinct from it. He wrote:

> Besides this natural means of selection, by which those individuals are preserved, whether in their egg, or larval, or mature state, which are best adapted to the place they fill in nature, there is a second agency at work in most unisexual animals tending to produce the same effect, namely, the struggle of the males for the females.[11]

The distinction between natural and sexual selection here was chiefly a logical distinction since they both tend "to produce the same effect." Sexual selection differed from natural selection in that it did not necessitate the death of individuals, but merely denied some an equal degree of reproductive success in relation to others. In 1858, Darwin's definition of sexual selection was really a description of natural selection in the reproductive context.

In the *Origin of Species*, the explanation of sexual selection was expanded somwhat. Darwin expressed the opinion that the struggle between males was not the only element in reproductive success, but that female animals choose their mates by the use of a primitive kind of aesthetic taste. In the first edition of the *Origin*, Darwin wrote:

> I can see no good reason to doubt that female birds, by selecting, during thousands of generations, the most melodious or beautiful males, according to their standard of beauty, might produce a marked effect.[12]

It was only in the *Descent of Man* in 1871 that Darwin was able to use all his collected notes on the subject of sexual selection, a subject which he declared in his autobiography, "had always greatly interested me."[13] Sexual selection was one of the few subjects he had been able to write about in full, employing all the materials he had collected.

Darwin frankly admitted that he had waited until many naturalists had begun to concur with his theory of evolution before venturing to air his theory with regard to "the descent of man." The work had been made easier for him by Lyell's *The Antiquity of Man* (1863), and Huxley's *Evidence* as to Man's Place in Nature.[14] Darwin's autobiographical memoir contains his stated feelings about the relation of a work on human evolution to the advancement of evolutionary theory. He wrote: "I was the more glad to do so [write on man], as it gave me an opportunity of fully discussing sexual selection."[15] Sexual selection was intimately connected in Darwin's thinking to the subjects of human origin, racial divergence, secondary sexual differences. The Descent of Man touched on almost every issue on which Darwin and Wallace disagreed: natural selection in animal and human evolution, sexual selection, utility as a criterion for selection, inheritance of acquired characters, and the validity of the analogy of domestic animals to wild species. Darwin acknowledged the situation a little sadly in a letter to Wallace: "Fate has ordained that almost every point on which we differ should be crowded into this vol."[16]

The meanings of colors to the animals themselves and the evolution of coloration were subjects often raised in the correspondence between the two naturalists, both while Wallace was in the Malay Archipelago and after his return to England in 1862. The subject of protective coloring and sexual selection was first alluded to by Darwin in his seventh letter to Wallace, of March 7, 1860, where he wrote:

> I fully admit the probability of "protective adaptation" having come into play with female butterflies as well as with female birds. I

> have good many facts which make me believe in sexual selection as applied to man, but whether I shall convince anyone else is very doubtful.[17]

The discussion lapsed until Wallace began to express his opinions on the evolution of man; they were first published in 1864 in the Anthropological Review.[18] In responding to Wallace's letter enclosing a copy of the paper on man, Darwin again underscored his opinion that the races of man have diverged by a "sort of sexual selection" based on different races having "a widely different standard of beauty."[19] In that same letter, Darwin generously offered Wallace his collection of notes on human origins. He had been through an extended period of illness and despaired of ever being well enough to pull the notes together into a book.

Naturally, Wallace was pleased to have received Darwin's invitation to use his materials, but delayed accepting on the grounds that he had already undertaken the "Narrative of my Travels" which eventually appeared as The Malay Archipelago.[20] In the course of his replying letter, Wallace defended his positions more fully on man. He ascribed any misunderstanding of his theory to his own personal failing of excessive haste in publishing. He realized that in his Anthropological Review article, he had emphasized the inadequacy of natural selection alone to produce an organism differing as greatly from its nearest extant relatives as man does from the other primate species. Also, Wallace protested, haste had caused him to neglect to point out that natural selection had not altogether ceased in its activity on men; for example, it may still act in respect to human colors. His studies of geographical distribution of races seemed to indicate that so much migration and displacement of human groups had taken place subsequent to the evolution of different characteristic racial colors that even in this trait natural selection's role in human evolution has been for a very long period secondary in its importance. It seemed to Wallace that the absence of either polygamy or choice of mates based on physical beauty denied the action of Darwin's theory of sexual selection in human beings. "The sexual selection you allude to will, also, I think have been equally uncertain in its results."[21]

Darwin's response shows that he was undaunted by Wallace's objections.

> Probably you are right on all the points you touch on except, as I think, about sexual selection, which I will not give up.
>
> My belief in it, however, is contingent on my general belief in sexual selection. It is an awful stretcher to believe that a peacock's tail was thus formed; but believing it, I believe in the same principle somewhat modified applied to man.[22]

It must have been clear to Darwin by June of 1864 that the notes he had collected on the evolution of man would not be of much use to Wallace. He withdrew his offer of them; it may also be that Darwin was feeling better and was again hopeful that he could undertake the "essay on man" himself.[23]

After the publication of Wallace's paper on the Malayan Papilionidae in 1866, the two men exchanged letters on the subject of the evolution of sexual dimorphism and polymorphism in species of Papilio.[24] Darwin was particularly interested in how new forms arise with respect to the laws of heredity. Wallace's attention was less on preservation of the traits as hereditary units than on the fact that they were preserved to serve utilitarian functions. Common data had quite a different meaning for the two correspondents.

The question of gaudy colors in caterpillars succeeded the discussion of Malayan Papilionidae. Darwin had come up against a puzzling problem (already alluded to above in Chapter IV). Why did caterpillars, which are not reproductively active, have the same kinds of brilliant colors Darwin believed had been preserved by sexual selection?[25] Wallace had answered that the conspicuous colors of caterpillars had been preserved as manifestation of unpalatability to their predators like that found in Bates's mimics and in Malayan Papilionidae.[26] Darwin was not to be won over to explanations based on protection just because caterpillar colors cannot be due to sexual selection.

As is so often the case in Darwin's correspondence with Wallace, Darwin accepted Wallace's statements without relinquishing any of his beliefs about sexual selection.

Darwin wrote in February 1867, replying to Wallace's remarks on coloration in larvae:

> I never heard anything more ingenious than your suggestion, and I hope you may be able to prove it true...It warms one's very blood to see a theory thus almost proved to be true. With respect to the beauty of male butterflies, I must as yet think that it is due to sexual selection...This being the case, the analogy of birds makes me believe in sexual selection with respect to colour in insects...
>
> The reason of my being so much interested just at present about sexual selection is that I have almost resolved to publish a little essay on the Origin of Mankind, and I still strongly think (though I failed to convince you, and this to me is the heaviest blow possible) that sexual selection has been the main agent in forming the races of man.[27]

Each naturalist should have liked to convince the other of the truth of his own position. Lacking that, however, Darwin and Wallace each were largely content to pursue his own subjects--Darwin following out his ideas on sexual selection, in both men and animals, and Wallace working on his theories of protective coloring in insects and birds. The amicable openness, generally so pleasant to both, was marred by a brief episode of irritation in April and May of 1867. Darwin again had the sense of being preceded by the younger naturalist.

The letter from Wallace which precipitated Darwin's irked response has, at last, been <u>partly</u> published by Peter Vorzimmer. Wallace wrote on April 26, 1867:

> I have lately hit upon a generalization connected with sexual characters which pleases me

> very much and I make no doubt will interest you. I have become more than ever convinced of the powerful effect of "protective resemblances" in determining and regulating the development of colour...I think this proves that the primary action of sexual selection is to produce colour pretty equally in both sexes, but that it is checked in the females by the immense importance of protection and the danger of conspicuous colouring.[28]

On the 29th, Darwin pointed out to Wallace that he (Darwin) had alluded to the subjects Wallace discussed in both the first and fourth editions of the *Origin*.[29] Furthermore, that questions directed to Wallace in person at the Zoological Society and the queries on brilliant colors in caterpillars were all a part of his work on the "Essay on Man."

Darwin knew perfectly well that Wallace's papers on animal coloring would be quite different from his own approach to coloring as resulting primarily from the action of sexual selection. Nevertheless, the claim to the territory was made with the following assertion. "I have collected all my old notes and partly written my discussion, and it would be flat work for me to give the leading idea as exclusively from you." In a post-script under his signature Darwin wrote, "It is curious how we hit on the same ideas."[30] It was no accident, of course. Both naturalists were proceeding to explore the logical ramifications of a general theory for special cases. Their perceptions of how the sexually dimorphic colors and other characters evolve was nevertheless quite different.

Wallace devoted most of his attention to the meanings of colors for the animal in adaptation. Darwin, on the contrary, centered his attention on characteristics considered useless in the general struggle for life, but useful in the struggle for mates.

Wallace was apparently astonished by Darwin's feelings regarding his plan to publish on the connections between animal color and nesting habits. He packed up his notes and sent them off to Down in the mail with a letter attempting to make the matter right with Darwin. Wallace wrote:

> I had thought of a short paper on "The Connection between the Colours of Female Birds and their Mode of Nidification"--but had rather leave it for you to treat as part of the really great subject of Sexual Selection--which combined with protective resemblances will I think when thoroughly worked out explain the whole colouring of the animal kingdom.[31]

Darwin returned Wallace's notes a little embarrassed at the circumstances which had motivated Wallace to send them. "I confess," Darwin wrote, "on receiving your note I felt rather flat at my work being almost thrown away, but I did not intend to show this feeling."[32] Darwin urged Wallace to continue his work on the sexual colors of birds and concluded his letter, "Forgive me, if you can, for a touch of illiberality about your paper."[33]

This exchange of letters is repeated here to sketch out the very significant differences between the two naturalists. They were very different personalities, but more significant for this study, they were quite dissimilar in their attitudes toward data. Wallace, in later years, was oriented toward general ideas and theories. Wilma George's title for her biography of Wallace, Biologist Philosopher, is absolutely correct, capturing the flavor of his approach to science.[34] He did not have Darwin's delight in the details of data-classes and, consequently, felt much less possessive of data. Knowing that he and Darwin were treating the phenomena very differently was enough for Wallace, but not for Darwin, who had a much clearer dedication to inductive methodology, at least when it came to his publications.

Ostensibly, disagreeing with Wallace caused Darwin more anxiety, although anxiety did not result in an acceptance of Wallace's point of view. In his April 15, 1868, letter to Wallace, Darwin wrote: "I am delighted to see that we really differ very little--not more than two men almost always will... I wish with all my heart I could go the whole length with you."[35] And again he stated on April 30, "But we shall never convince each other. I sometimes marvel at how truth progresses, so difficult is it for one man to convince another unless his mind is vacant."[36] Nearly every (published) letter in 1868 between the two naturalists discusses some aspect of sexual selection and animal coloring. Darwin's concern about reconciliation of his view and Wallace's was conspicuous.

> You will be pleased to hear that I am undergoing severe distress about protection and sexual selection: this morning I oscillated with joy towards you; this evening I have swung back to the old position, out of which I fear I shall never get.[37]

It comes as no surprise that Darwin never did get out of his old position. After all, he had developed his concept of sexual selection very nearly at the same time as his theory of evolution by natural selection. He had given it an augmenting proportion of his time and energy, especially after the _Variation of Animals and Plants under Domestication_ was in the press.

Both publically and privately, Wallace did his best to minimize the importance of their irreconciliable differences on the subject of sexual coloring in animals. On September 27, 1868, Wallace added a post-script to yet another letter which attempted to subsume sexual selection under natural selection: "Pray don't distress yourself on this subject. It will all come right in the end, and after all it is only an episode in your great work."[38]

James Marchant printed another letter to the same effect from the Wallace letter portfolio.

> I am sorry to find that our difference of opinion on this point is a source of anxiety to you.

> Pray do not let it be so. The truth will come out at last, and our difference may be the means of setting others to work who may set us both right.
>
> After all, this question is only an episode (though an important one) in the great question of the origin of species, and whether you or I are right will not at all affect the main doctrine--that is one comfort.
>
> I hope you will publish your treatise on Sexual Selection as a separate book as soon as possible, and then while you are going on with your other work, there will no doubt be found someone to battle with me over your facts, on this hard problem.[39]

For Wallace, the argumentation was enjoyable, win or lose. He was quite irrepressible in his optimism. Even in 1871 he was as ebullient as ever, writing once again to Darwin: "You will not be surprised to find that you have not convinced me on the 'female protection' question, but you _will_ be surprised to hear that I do not despair of convincing you."[40] Wallace seems to have entirely enjoyed the play of opposing ideas in the privacy of letters and found it amusing, but unlike Thomas Huxley, he took no particular delight in public controversy. He tried to decline to review Darwin's _Descent of Man_, since he would have felt obligated to criticise sexual selection in the _Academy_, a "highbrow" publication.[41] The editor prevailed upon Wallace to review the book by arguing that Wallace was capitulating to Darwin's wishes that Wallace not criticize the _Descent_.[42] When published, the review contained the expected dissent from sexual selection and urged the alternative explanation of dull coloration in females by natural selection for protection.[43]

The causes of the differences of opinion between Darwin and Wallace can only be speculated upon from their effects. For Darwin, the evolutionary continuum was complete from insects to man. He accepted the principle that the senses of animals and people were quantitatively different but qualitatively similar, that the laws of inheritance and the laws of evolutionary change were homogeneous throughout the organic world. Increasingly, Wallace came to doubt that natural processes were entirely uniform differing only in quantity rather than in qualities. Human beings introduced new problems of evolutionary change. But even in the lower organisms, it appeared that quite different senses resulted in differing perceptions negating the uniform principles Darwin required for sexual selection.

Darwin's theory of continuity of sexual selection from insects to man could be correct if, and only if, insects had similar perceptions of color to that of birds and mammals _and_ if they used their perceptions in the choice of mates. Neither seemed likely to Wallace, but the further ascription of an "aesthetic sense" which made "female choice" possible seemed absolutely gratuitous, lacking both observational confirmation and theoretical necessity. It seemed to him that ornithological studies indicated that female birds rarely even watch the inter-male rivalry.

In his autobiography, Wallace described his growing doubts that sexual selection involving female choice was any better in explaining the evolution of sex differences in birds than in butterflies.

> The first kind of sexual selection [combats between males] I hold as thoroughly as Darwin himself; the latter [choice of more musical or ornamental males by females] I at first accepted, following Darwin's conclusions from what appeared to be strong evidence explicable in no other way; but I soon came to doubt the possibility of such an explanation, at first from considering the fact that in butterflies sexual differences are as strongly marked as in birds, and it was to me impossible to accept female choice in their case, while, as the whole question of colour came to be better understood, I saw equally valid reasons for its total rejection even in birds and mammalia.[44]

Wallace believed that the choice of brightly colored males could almost never be subjected to observational testing. Female choice was an inference, not an observation; therefore, it must give way before empirical counter-evidence.

As his studies proceeded, Wallace found more and more different uses for color in birds other than those exclusively connected with reproduction. Many colors and patterns served male animals not so much to charm females as to intimidate other males. Furthermore, Wallace claimed that many of the bird colors Darwin ascribed to sexual selection served as recognition marks and were attributable to natural selection.

In insects more and more colors were accounted for as warning colors, not sexual in their function.

The Color Sense

Wallace's view of color as mostly protective in function had less appeal in the "scientific community" than Darwin's view of sexual selection had. Darwin's rather than Wallace's view was accepted by authors who searched in the lower animals for the aesthetic sensitivity which reached its highest state in human beings. Grant Allen, author of both scientific works and novels, wrote a book called The Colour Sense: Its Origin and Development. An Essay in Comparative Psychology (1879).[45] Allen tried to show that the evolution of sensitivity to color and delight in colors were both essentially continuous in the "scale of nature." He summarized his view in concluding:

> Insects produce flowers. Flowers produce the colour-sense in insects. The colour-sense produces a taste for colour. The taste for colour produces butterflies and brilliant beetles. Birds and mammals produce fruits. Fruits produce a taste for colour in birds and mammals. The taste for colour produces the external hues of humming-birds, parrots, and monkeys. Man's frugivorous ancestry produces in him a similar taste; and that taste produces the various results of human chromatic arts.[46]

From man's descent from the frugivorous quadrumana Allan moved with incredible swiftness to the British taste for the "golden haze which glimmers over the dreamy visions of

Turner!"[47]

Other naturalists who attempted to assess the presence of aesthetic judgment in animals and the qualities of their senses were more discriminating than Allan had been. George John Romanes accepted aesthetic taste in animals as Darwin had, but avoided Allan's anthropomorphism.[48] The better behavioral psychologists handled their discussions of both aethetic potentials and sexual display in animals with greater moderation. However, detailed observations of mating displays, territorial struggle, and threat were only worked out in the twentieth century. Psychologists and evolutionary theorists were forced, in the nineteenth century, to rely on the careful use of analogy extending their views from the few species which had been studied to those whose behaviors were less well known.

In her discussion of sexual selection, Wilma George correctly stated that the theory of sexual selection introduced an element of freedom of choice and of beauty into nature which appealed to some naturalists.[49] These two aspects, choice and beauty, are notably absent from the vision of nature under natural selection which, according to Tennyson, is "red in tooth and claw." It is interesting to speculate on whether the acceptance of sexual selection as a scientific theory was part of the revolt from a picture of nature ruled only by random variation and an inexorable and unmerciful natural selection. Clearly it was never Darwin's purpose to devise a theoretic wedge for the reinsertion of a benevolent view of nature's operation, in which insects and birds find pleasure, in a human sense, in their environments. Manufacturing world-pictures of maximal pleasure and minimal pain was a Utilitarian utopia which Darwin would not have promoted as describing the real world.

The Colour-Sense was reviewed in *Nature* by Wallace rather than by Darwin.[50] The book was criticized factually and theoretically. Why taste for colors should be aroused in animals under the influence of the colors of their foods and transferred to preferences in the colors of their mates Wallace could not fathom. In his opinion, a world without a general color-sense or without sexual selection involving female choice was not a world lacking in sublimity. Beauty in mimetic butterflies and in birds was not less beautiful if accounted for by natural selection in the struggle for life. If beauty is in the interpreting eye of the beholder, Wallace reasoned, it nevertheless requires only natural selection to form the eye and the capacity to interpret.

Utility was the principle to which Wallace repeatedly turned in his explanations of the evolution of colors and patterns in animals. He considered his devotion to utility among the hallmarks of his work if it were judged with respect to Darwin's. In *My Life*, Wallace wrote:

> The principle of 'utility,' which is one of its [evolutionary theory's] chief foundation-stones, I have always advocated unreservedly; while extending this principle to almost every kind and degree of coloration, and in maintaining the power of natural selection to increase the infertility of hybrid unions I have extended its range. Hence it is that some of my critics declare that I am more Darwinian than Darwin himself, and in this, I admit, they are not far wrong.[51]

Wallace had been criticized by Romanes and others on that very point, which only indicates that a critic's criticism may be another man's pride.

The principle of utility in connection with animal coloration is found in Wallace's _Contributions to the Theory of Natural Selection_ (1870), in _Tropical Nature_ (1878), in _Darwinism_ (1889) and more obliquely in _Malay Archipelago_ (1869). The collection of essays, _Contributions to the Theory of Natural Selection_, provided readers with reprints of Wallace's essays of 1855 and 1858 and with eight other essays displaying the range and depth of his contributions to evolutionary theory. Two essays were on mimicry, two on bird colors and nesting habits, two on human evolution, and two general treatments of evolution.

In one of the general chapters, "On Instinct in Man and Animals," Wallace voiced the same objections against instinct theorists he had launched against the "colour-sense."[52] Both theories violated the scientific rules of procedure. Neither instinct nor other occult senses should be introduced where ordinary faculties had not been ruled out positively. Wallace reminded readers of how little was known about the senses of animals, particularly the lower animals. Almost nothing was known about the range of hearing in insects, the wavelengths of light they could perceive, or the smells they could detect. Animals must be studied in order to ascertain their capacities for learning before resorting to explanations based on instinct. The senses and faculties of lower organisms were too little known to be discussed with certainty. However, the principle of parsimony dictates that no additional faculties ought to be posited where known senses were adequate to explain phenomena. It seemed to Wallace that many behaviors in animals were better explained as learned rather than as instinctive. Instinct theory and sexual dimorphism theory both erred in introducing ancillary explanations to supplement acknowledged mechanisms without showing the need for them. Making guesses about insect faculties on the basis of analogy was both immensely naive and counter-productive, Wallace stated:

> How can we pretend to fathom the profound mystery of their mental nature, and decide what, and how much, they can perceive or remember, reason or reflect! To leap at one bound from our own consciousness to that of an insect's, is as unreasonable and absurd as if, with a pretty good knowledge of the multiplication table, we were to go straight to the study of the calculus of function...In such a case would not error be inevitable, and would not continued study in the same direction only

> render the erroneous conclusions more in-grained and more irremovable.[53]

Wallace opposed both instinct and sexual selection on the ground that they each presupposed unproved continuity in senses from insect to man. Except for natural selection, Wallace did not find such ambience in theories compelling.

Coloration by Law

In Wallace's other general essay on evolution, "Creation by Law," he also discussed animal colors as part of his review of the Duke of Argyll's book The Reign of Law.[54] The Reign of Law presented the view that the Creator created beautiful objects in the world for his own pleasure. Beauty had been introduced in many cases for beauty's own sake and not for any utilitarian purpose, the Duke wrote, although there were many cases in which animals and plants were both beautiful and useful. The humming-birds were the Duke of Argyll's examples of beauty without utility.

> Mere ornament and variety of form, and these for their own sake, is the only principle or rule with reference to which Creative Power seems to have worked in these wonderful and beautiful birds...A crest of topaz is no better in the struggle for existence than a crest of sapphire...A tail is not affected for the purposes of flight, whether its marginal or its central feathers are decorated with white... Mere beauty and mere variety, for their own sake, are objects which we ourselves seek when we can make the Forces of Nature subordinate to the attainment of them. There seems no conceivable reason why we should doubt or

> question, that these are ends and aims also in the forms given to living organisms.[55]

Wallace's retort to the Duke's design argument and analogy based on human tastes must have pleased Darwin. Not only did Wallace strongly oppose the metaphysical arguments of *The Reign of Law*, but he also ascribed the humming-bird's jewelled colors and crests to sexual selection:

> Experiment and observation have shown, that this kind of sexual selection does actually take place; and the laws of inheritance would necessarily lead to the further development of any individual peculiarity that was attractive, and thus the splendour of the humming-birds is directly connected with their very existence.[56]

In later writings on the same subject, for example in *Darwinism* in 1889, the colors of humming-bird crests and tails were explained as recognition marks, not as the result of sexual selection. The humming-birds' characteristic marking had been produced along with: "Structural differences on which the differentiation of species depends."[57] Conspicuous color differences served to separate closely related species. A topaz-crested humming-bird and a sapphire-crested one might be members of the same genus; hybridization is minimized when clear color distinctions are easily seen by the birds.

The discussion of birds in the *Contributions* brought Wallace as close to Darwin's opinion as he was ever to come, but even here there were significant differences. The two chapters, "The Philosophy of Birds' Nests" and "A Theory of Birds' Nests: Shewing the Relation of Certain Sexual Differences of Colour in Birds To Their Mode of Nidification," were Wallace's forays into the combined study of bird behavior and evolutionary theory.[58] He was convinced that bird behavior was less rigid and stereotyped than most naturalists believed them to be. Birds could change nest-building practices adaptively when given inducement and opportunity to innovate. The failure of caged birds to make proper nests even when given the appropriate building materials seemed to indicate that they had not had opportunity to learn nest construction from being themselves reared in a nest. Instinct theorists used that same information to argue oppositely--that captive birds fail to nest because their instincts were not activated or were adversely affected by captivity.

The "philosophy" and "theory" of birds' nests were parts of Wallace's alternative understanding of the evolution of colors in birds, especially sexually dimorphic colors. Wallace devoted considerable effort to a survey of birds in which he tried to show a correlation between nest type and sexual dimorphism. The connection of different modes of life to different degrees of color dimorphism would support natural selection as the major agency of evolution, against Darwin's sexual selection theory including female choice.

The data seemed to support a general rule which Wallace presented in italics:

> It is the rule--*that when both sexes are of strikingly gay and conspicuous colours, the nest is of the second class, or such as to conceal the sitting bird; while, whenever there is a striking contrast of colours, the male being gay and conspicuous, the female dull and obscure, the nest is open and the sitting bird exposed to view.*[59]

Even aberrent bird forms confirmed his position. The rare cases in which the female bird was more brilliantly colored than the male were those cases in which the male rather than the female incubates the eggs.[60] Wallace freely acknowledged that his belief in natural selection rather than sexual selection as the major evolutionary cause of coloration in birds was based on his study of butterflies. For Wallace protective, obscure colors were not "primitive," or "primordial" but evolved by the elimination of bright color from the exposed bird sitting on the nest. The Papilionidae showed no innate tendency to greater color development in male than in female butterflies, since many mimics displaying bright colors were limited to the female sex. Females are not dull because they are females, but because they are eliminated if they are not. The evidence provided by female mimicry in butterflies was supplied by brilliantly colored female birds which live in holes and domed nests. Wallace linked the birds' nest studies to the work on butterflies:

> I may mention, as a sufficient proof, that these are not isolated facts, that I was first

> led to see their relation to each other by the study of an analogous though distinct set of phenomena among insects, that of protective resemblance and "mimicry."[61]

Whenever bright colors are not disadvantageous to the individuals having them, Wallace believed that both sexes would tend to develop them equally. Using Darwin's term "sexual selection" in a special sense Wallace asserted:

> The normal action of "sexual selection" is to develop colour and beauty in both sexes, by the preservation and multiplication of all varieties of colour in either sex which are pleasing to the other.[62]

Equal intensity of colors in both sexes is to be expected except in cases where it is suppressed by natural selection. Whatever the colors of a bird or insect are, they are the result of a long evolutionary course in which the proper degree of cryptic color, in relation to conspicuous color, is worked out in each sex of a species.

The means by which each sex could pursue an independent evolutionary course was immensely puzzling. Neither Wallace nor Darwin had a very clear understanding of how the laws of inheritance worked to render the sexes so different. Of the two, Darwin devoted considerably more effort to unravelling the tangle of questions pertaining to heredity.[63] Wallace devised no theories comparable to Darwin's theory of pangenesis.

Darwin supposed that bright colors arose, almost always, in the more innovative male sex. Generally, secondary sexual traits were passed from father to son or from mother to daughter. In some cases, Darwin expalined, sexual colors could be inherited from the male parent by offspring of both sexes. Darwin could see that Wallace's studies of birds' nests indicated that bright colors are not associated only with "maleness" per se. Nevertheless, in Darwin's theory, bright colors and patterns were generally evolved from duller protective colors. Thus, the female was the sex nearer the ancestral colors.[64] Sexual selection operating by a combination of inter-male combat and female aesthetic taste produced the dimorphism in color and patterns of the two sexes.

One of Darwin's chief objections to Wallace's theory was that it offered no clear explanation of how the laws of inheritance worked. Darwin's criticisms were well founded, since Wallace was working almost exclusively on the preservation of variations rather than on how they are transmitted. Wallace's faith in the efficacy of natural selection to produce protective adaptations in one sex or in both sexes of a species was not supported by any elaborate "genetic" theory. Wallace's pragmatic approach to questions of heredity could hardly fail to be criticized. His understanding of birds was mainly based on what could be seen rather than on what was understood theoretically. The analogy to mimicry studies is very clear in Wallace's work on birds.

Natural selection had been adequate to explain mimetic polymorphism in butterflies and Wallace could not see why it should not be adequate with respect to birds. In "Mimicry, and Other Protective Resemblances Among Animals," Wallace explained why natural selection would have most impact on altering female coloring.

> At a longer distance it might be mistaken for one of the uneatable group, and so be passed by and gain another day's life, which might in many cases be sufficient for it to lay a quantity of eggs and leave a numerous progeny, many of which would inherit the peculiarity which had been the safeguard of their parent.[65]

Questions of adaptive advantages were more important in Wallace's theory than hereditary mechanisms.

Two noted anti-evolutionists were quick to point out shortcomings in Wallace's theory of bird colors and nests with respect to what they revealed about adaptation. The Duke of Argyll and Andrew Murray published rebuttals of Wallace's theory in the same issue of the Journal of Travel and Natural History which carried Wallace's paper in 1869.[66] Arguing in a similar vein as in the Reign of Law, the Duke of Argyll claimed that adaptive coloration proved the activity of "Mind directing the forces of Nature in a preconceived and preordained direction."[67] Furthermore, the Duke doubted that domed bird houses were less conspicuous to a predator than a smaller open nest. As a proponent of design, the Duke could freely admit that Wallace had shown that bright colors were not exclusively a male prerogative. He could allow Wallace his fact, but not his inference.

> All, therefore, that can be said is, that it is one of those special adaptations for special purposes which are so abundant in nature, but which leave us absolutely in the dark as to the physical agencies by which they are secured.[68]

Wallace's theory of birds' nests had a quality of "elasticity" which his opponents found immensely galling.[69] It seemed to be designed to embrace any fact whatsoever. Wallace's assumption that the evolution of sexual dimorphism is only a subset of phenomena which constitute one aspect of the evolution of species from some antecedent species was not acceptable to the Duke of Argyll.

Andrew Murray's article against the theory of birds' nests demonstrates the long-lasting inability of some naturalists to understand what the Darwinians meant when they used their characteristic phrases, for example, natural selection, gradual change, and struggle for life. Murray believed that nature unerringly produced the fittest: "She never errs and never wastes."[70] For Murray, natural laws were so constituted that there needed to be no natural selection because there are no random variations for it to select. The frequency in which animal colors duplicate the colors of the environment illustrated to Murray's satisfaction the action of a law embracing all of nature.

> Every artist, every admirer of scenery is

> familiar with the harmony of colour in nature. We do not believe that this is fortuitous. It seems as if there were a polarization of colour as there is of magnetism and electricity, which reduces everything into harmony.[71]

This electro-magnetic theory of harmony between animals, plants and minerals was little more than a naturalistic figleaf for design. Murray stated that recent embryological studies showed that vertebrate embryos have electrical potentials confirming the opinion that electricity and magnetism are everywhere working homogeneously in nature. Murray concluded with an allusion to the psalms and with a denial that the Darwinians had contributed to the understanding of animal coloration.

> We are fearfully and wonderfully made, and there are many things in the making of which we have as yet no idea--the distribution of colour is one of them.[72]

Murray had made no major change in his position since "On the Disguises of Nature" of 1860.[73]

Wallace's papers on bird nests, sexual differences, and natural selection met little contemporary praise. The anti-evolutionists opposed his ideas and Darwin, also, wondered if Wallace had not confused the cause with the result. How could he be so sure that it was not cryptically colored birds which could use open nests rather than that they evolved cryptic colors because they built open nests? The question is still debated in the twentieth century. The general correspondence between color in female birds and nest type which Wallace alleged has been upheld by modern ornithologists, at least for Malayan and British birds.[74]

Descent of Organisms

<u>The Contributions to the Theory of Natural Selection</u> was published only a little earlier than the <u>Descent of Man</u>, and both books include similar data. Both works discuss insects, birds, and people with respect to their evolution. In each, both the authors grappled with problems of animal behavior, particularly behaviors associated with courtship and reproduction. Of the two, Darwin was the more willing to accept the inheritance of habit. Instincts, he thought, like physical traits are subject to variation and to preservation of adaptive variations. Also, evidence for variability of innate behavior is provided by the loss of instinctive behaviors observed in domestic animals.

In Darwin's theory, instinct was an element common to animals; it was often displayed in conjunction with learned behaviors. Distinct mental actions are often combined in animals. Darwin summarized his opinion in the sixth edition of the Origin: "A little dose, as Pierre Huber expresses it, of judgment or reason, often comes into play, even in animals very low in the scale of nature."[75] The judgment that animal behavior was rarely divisible into two neat categories, instinctive or learned, is reflected in Darwin's reluctance to fix a definition of instinct except in connection with particular behavior. Since Darwin allowed for the inheritance of habit in his theory of pangenesis, he had no <u>a priori</u> objection, either scientific or philosophical in nature, to instincts being one factor in animal and human behavior. The gregarious tendencies of organisms, for example, "may be attributed in part to [inherited] habit, but chiefly to natural selection."[76]

In assessing Darwin's view of behavior as a mixture of learned and instinctive components, Michael Ghiselin concluded:

> To attribute "higher" mental processes to "lower" organisms was one way of arguing for evolution. Yet we may see that he [Darwin] is not exalting the mental capacities of worms, but is seeking to degrade the "higher" animals.[77]

In Darwin's theory, the mixture shifts from relatively stereotyped behavior systems in lower animals to less rigidly programmed systems in higher animals in which reason plays more of a part. Throughout, however, Darwin's view places the accent on continuity rather than on the differences between the levels of organization.

Ghiselin's contention that Darwin preferred to see higher mental functions in lower animals in order to degrade higher animals seems fundamentally correct in the context of the whole Darwinian *corpus*. Julian Huxley came to a similar view from his reading of Darwin's work. Even the naming of the *Descent of Man* indicates the firmness with which Darwin underscores the connections of human history and animal ancestry. Huxley wrote:

> Darwin's primary and main aim was to provide convincing evidence that organisms were not immutable creations but had evolved by natural means from something different; and this implied a focussing of attention on their past history. This preoccupation of his with origins is revealed in the titles he chose for his two greatest works--*The Origin of Species* and *The Descent of Man*--though *The Evolution of Organisms* and *The Ascent of Man* would in fact have been more appropriate.[78]

Huxley's and Ghiselin's points are mutually confirming. Both agree that Darwin sought the origin of the higher in the lower. Higher species descend from lower, or at least earlier, ones, and the attributes of later forms are often found in the earlier. When Darwin likened worm behavior to human behavior, mingling learning and instinct, and when he ascribed sexual selection, including aesthetic judgment, to insects, it is more appropriate to describe him as *zoomorphic* with respect to man than *anthropomorphic* with respect to insects.

The *Descent of Man* was, by far, Darwin's lengthiest expostion of sexual selection, although it was not his last. After the *Descent*, he published three papers on sexual selection in particular groups, "Sexual Selection in Relation to

Monkeys" and "The Sexual Colours of Certain Butterflies" appeared in Nature in 1876 and 1880.[79] Finally, a paper was read before the Zoological Society April 18, 1882, "On the Modification of a Race of Syrian Street-dogs by means of Sexual Selection," by W. Van Dyck. With a preliminary notice by Charles Darwin."[80] Interestingly, it was the last publication in Darwin's lifetime. He died the following day, April 19, 1882.

Categories of Coloration

No book-length work on animal coloration as a separate topic related to evolutionary theory had been published before Darwin's death. The first person to publish such a book was Alfred Tylor, whose Colouration in Animals and Plants was published in 1886.[81] His work follows in the form established by the Rev. H. H. Higgins.[82] Tylor, like Higgins, had every hope of decoding nature and proceeded to do so in even a more eccentric fashion. Tylor was an evolutionist like Samuel Butler rather than like Darwin or Wallace, and he accepted Butler's concept of inherited memory.[83] It provided him with a linear view of change and avoided confrontation with adverse variations or incomplete adaptation. "The moral of all this is that things become easy by repetition; that without experience nothing can be done well, and that the course of development is always in one direction, because the memory of the road traversed is not forgotten."[84] On the basis of community of descent, Tylor, following Butler, espoused the belief that higher animals are beneficiaries of longer collective experience.

> The fundamental law of decoration is a structural one. Assuming, as we do, that memory has played a most important part in evolution, it follows that all living matter has a profound experience in decorating its organs--it is knowledge just as anciently acquired, and as perfectly, as the power of digestion. This colour was produced under the influence of light--so it is even in opaque animals...We might expect [that] even in opaque animals the colouring would still follow structural lines, and there should still be traces of this...This is precisely what we do find.[85]

According to Tylor, two great laws of animal coloring have been found. The first, the Law of Exposure, depends on the action of light and accounts for the multitude of cases in which the dorsal sides of animals are darker than the ventral. The second law, the Law of Structure, was perhaps the most interesting aspect of Tylor's book.[86] The Law of Structure accounted for the coloration on the surface of an animal that seemed to sketch out underlying structures. The observation that the spinal stripes of many mammals are frequently interrupted was interpreted as caused by its composition out of separate vertebrae. The tiger, therefore, was said to have its cervical vertebrae indicated by seven stripes. Nerves also were often indicated by dark lines in the cat family, in the vertebrates in butterflies among the invertebrates.

Although Alfred Tylor has the distinction of being the first to write a book on the coloration of animals and plants in connection with evolution, it was flawed by both theoretical shortcomings and by a lack of knowledge of the rapidly growing literature. Tylor had by no means mastered the studies by August Weismann to which he alluded.

If the study of animal coloration began inauspiciously, it was not lacking in interest or enthusiasm. In the 1880's the debates about mimicry were in full cry. During the period, it was Wallace who did most to organize the study of animal color and evolution. In his 1877 paper on "The Colors of Animals and Plants," Wallace presented a list of functions of color which he later elaborated. In *Macmillan's Magazine* and in *The American Naturalist*, which also printed the article in 1877, Wallace presented a table listing the various functions of color.[88]

Animals	1. Protective colors	
	2. Warning colors	a. of creatures specially protected. b. of defenseless creatures, mimicking a.
	3. Sexual colors	
	4. Typical colors	
Plants	5. Attractive colors	

The first class, protective colors, did not require much additional comment in 1877 above a brief recapitulation of his position on cryptic, camouflage coloration. The second class, also, had been treated before in the paper on the

Malayan Papilionidae, published in 1865 and again in his paper on mimicry in 1867. The discussion of sexual colors was also more compactly discussed in the 1877 paper compared to its presentation elsewhere. The impression Wallace's chapter leaves is of a wide-ranging theory encompassing the diversity of natural pehnomena. Wallace repeated for the readers of Macmillans Magazine and the American Naturalist the points he had made against Darwin's view of sexual selection many times before. Sexual dimorphism is the product of natural selection supplemented by the correlation of bright colors and general healthiness. In the rivalry between males the individual with the most beautiful plumage generally wins, not because he is beautiful, but because he is the better animal. The exercise of choice by females on the basis of plumage is unconfirmed and unlikely.

> The greater intensity of coloration in the male which may be termed the normal sexual difference--would be further developed by the combats of the males for the possession of the females. The most vigorous and energetic usually being able to rear the most offspring, intensity of color if dependent on or correlate with vigor, would tend to increase.[90]

Since healthy, vigorous animals are also brilliantly colored and pass those traits on to their offspring, natural and sexual selection work to the same effect, preserving and intensifying colors. Wallace, also, took the opportunity to repeat his opinion that color appears equally in the two sexes but is suppressed from developing in female animals by natural selection.

Under the rubric "sexual colors," Wallace discussed phenomena Darwin considered to be sexual in their function even though Wallace disagreed with the designation. He found two new functions for color which diminished the category of sexual colors--recognition and deflexion colors. Illustrating the category of recognition colors, Wallace explained the rabbit's white tail as a signal flag which the young could recognize and follow, rather than expalining it as Darwin had as a failure of protective coloring, probably due to sexual selection.[91] Also, the eye-like spots, the ocelli, and patches of color at the tips and margins of butterfly wings were described in the same paper as protective adaptations rather than as exclusively sexual in their uses by the animal.[92]

In the _Descent of Man_, Darwin explained the bright patches of the orange-tip butterfly as sexual coloration, since the underside was protectively colored.

> In our English orange-tip, the under surface resembles the flower-head of the wild parsely, on which the butterfly often rests at night. The same reason which compels us to believe that the lower surfaces have here been coloured for the sake of protection, leads us to deny that the wings have been tipped with bright orange for the same purpose, especially when this character is confined to the males.[93]

It seemed to Wallace that bright patches of color, placed at a maximum distance from the vital organs of the butterfly, might well serve to divert the attacks of birds, allowing the insect to escape, its wing slightly tattered perhaps but with its life.

The category of "typical" colors contained an assortment of colors not discussed under the other headings. These were colors which seemed to Wallace to appear spontaneously in animals and were not suppressed by natural selection nor retained to function in camouflage or recognition. Typical colors were thought to be produced as a normal by-product of development and metabolism--coloration being the most intense where the surplus of vital energy was greatest.[94] In Tropical Nature, "typical colors" were renamed "normal colors" and after the publication of Darwinism, Wallace added a footnote to the 1891 edition of Tropical Nature: "Many, or perhaps all, of these are now believed to be diversely coloured for purposes of recognition."[95]

The Origin of Species and Darwinism

Wallace's Darwinism was his final and fullest expression of views on animal coloration and their significance for the Darwinian theory. In the preface to Darwinism, Wallace indicated that he had tried in the work to strengthen the weaknesses of Darwin's own work.[96] He had avoided the use of the analogy between domestic and wild species. Furthermore, he had presented the subjects in a different sequence. Darwin's work was hard for many people to read, with its presentation of two chapters on variation before coming to the fundamental issue of "The Struggle for Existence." The parallel between _Darwinism_ and the _Origin of Species_ is nonetheless remarkable, despite Wallace's alterations. The second half of _Darwinism_ deals primarily with questions raised in the _Descent of Man_. Gerald Henderson remarked on the similarity of the organization of the two books in his doctoral thesis.[97] The parallel given below contains major alterations of Henderson's comparison, although it was suggested by his work.

The Origin of Species	_Darwinism_
An Historical Sketch and Introduction	I What are "Species," and What is meant by their "Origin"
I Variation under Domestication	II The Struggle for Existence
II Variation under Nature	III The Variability of Species in a State of Nature
III Struggle for Existence	IV Variation of Domesticated Animals and Cultivated Plants
IV Natural Selection: or, the Survival of the Fittest	V Natural Selection by Variation and Survival of the Fittest
V Laws of Variation	
VI Difficulties of the Theory	VI Difficulties and Objections
VII Miscellaneous Objections to the Theory of Natural Selection	

	The Origin of Species		Darwinism
VIII	Instinct		
IX	Hybridism	VII	On the Infertility of Crosses between Distinct Species and the Usual Sterility of their Hybrid Offspring
		VIII	The Origin and Uses of Color in Animals
		IX	Warning Coloration and Mimicry
	DESCENT OF MAN, PART II	X	Colors and Ornaments
		XI	The Special Colors of Plants: Their Origin and Purpose
X	On the Imperfection of the Geological Record	XIII	The Geological Evidences of Evolution
XI	On the Geological Succession of Organic Beings		
XII	Geographical Distribution	XII	The Geographical Distribution of Organisms
XIII	Geographical Diestribution (continued)		
XIV	Mutual Affinities of Organic Beings: Morphological: Embryological: Rudimentary Organs	XIV	Fundamental Problems in Relation to Variation and Heredity
	DESCENT OF MAN, PART I	XV	Darwinism Applied to Man
XV	Recapitulation and Conclusion		(part of chapter XV)

Largely motivated by persistant objection to natural selection and the success of his attempts to explain selection on his tour of America, Wallace had decided that a book vindicating his theory would be in order. Remembering the year 1887, he wrote:

> Many of my correspondents, as well as persons I met in America, told me that they could not understand Darwin's "Origin of Species," but they did understand my lecture on "Darwinism"; and it therefore occurred to me that a popular exposition of the subject might be useful, not only as enabling the general reader to understand Darwin, but also to serve as an answer to the many articles and books professiong to disprove the theory of natural selection. During the whole of the year 1888 I was engaged in writing this book, which though largely following the lines of Darwin's work, contained a great many new features, and dwelt especially with those parts of the subject which had been most generally misunderstood.[98]

Wallace believed, although he did not like to say so, that some of the misunderstandings of "Darwinism" had been occasioned by Darwin himself. Wallace claimed that his work presented readers with Darwin's early views from which he had receded in later years because of objections to his theory. Asserting that if the objections could be proven unsound, the need to retract the province of natural selection was thus avoided, Wallace wrote: "I insist on the greater efficacy of natural selection. This is pre-eminently the Darwinian doctrine, and I therefore claim for my book the position of being the advocate of pure Darwinism."[99] That was the only honor Wallace asked; he went on to liken Darwin's achievements to Newton's--surely the highest of British accolades.[100]

Besides recounting his own work on animal color for the new book, Wallace had to assimilate a truly burgeoning quantity of new literature from both observation and experimentation. The general theory of coloration had rapidly become more complex; typical colors were displaced by recognition colors, and protective colors required two new subdivisions, that is, terrorizing and alluring colors. Plant coloration, also, was treated at greater length and in greater detail in Darwinism. There Wallace set out recent findings of cryptic colors in plants, mimicry of unpalatable fungi, and attractive (alluring) colors to affect fertilization.[101]

Mimicry theory also was given a final assessment. Mimicry theory had been altered repeatedly since Bates's paper on the Heliconiidae in 1862. Wallace had played a very great role in the extension of the theory both theoretically and empirically. His pioneer study of mimetic polymorphism in Malayan Papilionidae is a classic in entomological literature. Wallace, unlike Bates, welcomed Müller's discovery of mimicry in species of different genera of unpalatable butterflies. In _Darwinism_, Wallace applied the concept of Müllerian mimicry to cases of similar color and pattern among the species of a single unpalatable genus.[102] From originally very simple beginnings, Wallace's theory of warning coloration had grown to include many phenomena. Besides true warnings of unpalatability and stings and the general category of mimetic colors, Wallace could include the explanation of bright colors in larval forms, a few cases of mimicry in plants, and mimicry in vertebrates. Mimicry in snakes, birds, and mammals had been described in the _Contributions_, but in 1889 they were part of an increasingly general theory.

Just less than one-third of the volume of _Darwinism_ is devoted to Wallace's theory of color in animals and plants. It was, in his personal opinion, one of his major accomplishments as an evolutionary biologist. The lectures in America on animal colors had been a successful vehicle for explaining the concept of natural selection to audiences of working people; he hoped for the same result in a work for the "general reader." In order to employ animal coloration in defense of natural selection, it is ironic that Wallace felt that to do so he must deny Darwin's explanation of the same data.

Wallace did not expect that _Darwinism_ would present the last word on the relation between Darwinian theory and animal coloration. In fact he did all within his power to see that it did not. He encouraged younger evolutionists in their researches. Edward B. Poulton was Wallace's immediate successor on the subject of animal colors and evolution.[103] Although he used and benefited from Wallace's advice in letters and his published work, Poulton carried the study of animal coloration far beyond the beginnings Wallace had made.

Edward Poulton was a biologist in contrast to a naturalist. Poulton was a skilled experimentalist who brought laboratory techniques to bear on questions in organism-level zoology. His book The Colors of Animals is perhaps justly more frequently cited as the beginning study of adaptive coloring by a Darwinian.[104] Poulton, who lived until 1943, advocated natural selection during the long "eclipse" of Darwinism when Neo-Lamarckian theory and mutations-theorie were in the ascendency over "pure" Darwinism among biologists. One good service Poulton performed for the study of adaptive coloration in animals was the rigorous use of exact terms. In collaboration with a friend, Arthur Sidgwick, Poulton chose Greek-based terms for the uses of color which have ever since made distinctions easier and sharper between the various biological functions of color in animals.[105] The following tabulation appeared in his book:

<table>
<tr><td colspan="2">I. Apatetic Colours[106]
Colours resembling some part of the environment or the appearance of another species.</td><td rowspan="2">II. Sematic Colours.--Warning & Signalling Colours.</td><td rowspan="3">III. Epigamic Colours.--Colours Displayed in Courtship.</td></tr>
<tr><td>A. Cryptic Colours--Protective and Aggressive Resemblances</td><td>B. Pseudosematic Colours--False Warning and Signalling Colours</td></tr>
<tr><td>1. Procryptic Colours--Protective Resemblances</td><td>1. Pseudaposematic Colours.--Protective Mimicry</td><td>1. Aposematic Colours. Warning Colours.</td></tr>
</table>

[Apatetic Colours]		[Sematic Colours]	[Epigamic Colours]
2. Anticryptic Colours.--Aggressive Resemblances	2. Pseudepisematic Colours.--Aggressive Mimicry and Alluring Colouration	2. Episematic Colours.--Recognition Markings	

Some authors, like Hugh Cott, prefer the English version of Poulton's terms, for example "alluring coloration" to pseudepisematic coloration.[107] They are still the categories of the uses of color that Wallace devised and that Poulton refined which have maintained their value for present-day researchers.

Poulton's discussion of Apatetic and Sematic Colors is largely repetitions with amplifications of Wallace's work. Only on Epigamic Colors does *The Colours of Animals* differ much with *Darwinism*. Poulton found Darwin's theory of sexual selection, including a female aethetic sense, more compelling than Wallace's alternatives and objections to it. The metaphor with which Poulton explained sexual selection to the reader was extremely clear and placed sexual selection in perspective as an ally of natural selection instead of its opponent. He wrote:

> Natural Selection is a qualifying examination which must be passed by all candidates for honours: Sexual Selection is an honours examination, in which many who have passed the previous examinations will be rejected.

> But the conditions for qualifying are more rigid than in any existing system; for the candidates who have barely qualified, or have qualified by some piece of luck, or have failed to keep up to the necessary standard in after life, will in the end be excluded from the advantages of any honours they may have gained.[108]

Thus, natural selection makes certain that males which are reproductively successful on account of their beauty are also among the most healthy, vigorous specimens. Its action is complemented by that of sexual selection which maintains beauty in the offspring of the fittest animals. This account of sexual selection was similar to Wallace's view; but Poulton went beyond the "honours examination" metaphor and accepted Darwin's opinion that display on the part of rival males before females strongly implies choice on the basis of display. In 1890, the evidence for choice simply outweighed the evidence against it. The degree to which female choice enters into the concept of sexual selection is still a matter of study by entomologists and ornithologists. Julian Huxley emphasized several aspects of bird behavior played down or omitted in nineteenth-century studies of the uses of color; two examples are display rituals minimizing actual combat, and mutual display and selection.[109] As confidence grew in natural selection and as more emphasis began to be placed on inter-male rivalry, the importance of female choice in selection theory has declined. In the period of

Wallace's controversy with Darwin, that confidence in natural selection was by no means common. Points of theory now considered minor then seemed of major significance in proving Darwinism right or wrong.

Animal Colors and Evolutionary Biology

In fact, animal coloration was a kind of "ideological football" in the 1890's and in the early twentieth century. People used the data of animal coloring to test various evolutionary theories. August Weismann's Studies in the Theory of Descent, 1882, presented a wealth of detailed observations on individual variations in caterpillar color and markings.[110] Weismann was presenting as persuasively as he could the possibility of evolution occurring on the basis of selection of individual variation alone. The Evolution Theory of 1904 was a much more general work and similar to Wallace's Darwinism in having chapters on (1) coloration of animals in connection with selection, (2) mimicry, (3) sexual selection, and (4) protective adaptations in plants.[111] The work of Wallace, Poulton, and Weismann, separately and collectively, evoked sharp retorts from mutation theoriests and neo-Lamarckians, and made the study of animal color a crucial test of evolutionary theory.

In advocating mutations as the vera causa of evolution, William Bateson included a long section of the evolution of color markings in Lepidoptera in his _Materials for the Study of Variation Treated with Especial Regard to Discontinuity in the Origin of Species_, 1894.[112] Similarly, Gustav Heinrich Eimer in 1890 and Alpheus Packard in 1904 took up the subject of animal colors from the neo-Lamarckian perspective.[113] Some authors of works on the evolution of color and patterns were merely ambivalent about which theory was correct. It seemed to some that nothing very meaningful could be concluded about the evolution of color and pattern. Marion T. Newbigin doubted that natural selection was the most important factor in the evolution of color, but admitted that in 1898 no other truly cogent alternative had been developed. An equally uncertain approach was taken by Frank Beddard _in Animal Colouration, an Account of the Principal Facts and Theories Relating to the Colours and Markings of Animals_.[115] Beddard tended to believe that the advocates of natural selection held a very weak position for explaining the evolution of protective coloration. He stated:

> It is quite unphilosophical to assume without further inquiry the general action of natural selection in producing, or at least intensifying, the similarity of colour between an animal and its usual environment. The _onus probandi_ lies with those who advocate the action of natural selection. The usefulness of each case of colour resemblance must be demonstrated; and any facts which tend to prove that it is of no particular use have far greater weight against the theory than suggestion, merely based upon human ideas of resemblance and difference, have in its favour.[116]

The advocates of Darwinian evolution were used to having the onus of proof on their side. It had never, really, been otherwise. Wallace was fully justified in his exclamation to Poulton in 1904:

> What a miserable abortion of a theory is "Mutation," which the Americans [referring to T. H. Morgan and his book Evolution and Adaptation, 1904] now seem to be taking up in place of Lamarckism, "superceded." Anything rather than Darwinism![117]

It seemed to the Darwinians that whenever solid evidence of variation and selection were found the anti-Darwinian forces unerringly found ways to shift the grounds of discussion. A general acceptance of Darwinian selection was attained only in the 1930's and '40's through the efforts of workers who combined current scholarship in genetics, population biology and evolutionary theory.

Among the names of those who participated in the "evolutionary synthesis" are several who used data on animal coloration to effect the synthesis. Animal coloration has continued to provide Darwinism with some of its best data. The work of G. D. Hale Carpenter, E. B. Ford, and E. B. Poulton in insects and of J. Huxley and E. Mayr in birds and of R. A. Fisher in population genetics has contributed to the modern understanding of the genetics of evolutionary change, with respect to adaptive coloration in animals.[118]

Important advances in pigment chemistry and in animal

behavior have altered modern understanding of the biological function of color since the nineteenth century. Only in the twentieth century has it been clear how importantly interconnected reproductive physiology and behavior are. For example, male courtship behavior is now known to function in some birds to synchronize the reproductive rhythms of the two sexes which maximizes the reproductive success of the mating.[119] David Lack's and Konrad Lorenz's studies on aggressive display have added to the understanding of the uses of display an element not developed in Wallace or Darwin's writing.[120] Much greater reliance is now placed on the uses of color as recognition signals, amplifying Wallace's concept far beyond his own conception of the range of uses recognition marks have. His opinion that color is produced as an expression of abundant vitality, however, is no longer held by modern biologists.

The importance of Darwin's and Wallace's contributions to the combined subjects of adaptive coloration and evolutionary theory does not really derive from the current favor in which their theories are held. Their work provided the foundation-stones for an edifice of research which continues to be added to in current periodical literature. Even the differences between the two original Darwinians served to generate interest in the subject of adaptive coloration as a testing topic for theories. It is fascinating that the modern researchers allege "Darwin's missing evidence" has been found in the industrial melanism of moths.[121] The "proof" of natural selection is still sought in the predatory relationships of birds on Lepidoptera graphically showing natural selection at work. Differential survival still produces a population with adaptive coloration as much in the trees near sooty English cities in the 1950's as in the Amazonian Forests where Bates found mimetic butterflies a century earlier.

The controversy between Darwin and Wallace on the nature of selection of secondary sexual characters in animals lasted beyond the life-span of either participant in the controversy. Some elements of their inability to agree devolved from the repertory of analogies and ways of thinking unique to each individual. Darwin's understanding of animal coloration was in large measure dependent on the appropriateness of the domestic analogy to wild species. Wallace began from the opposite point of view, with insects and birds in the "state of nature." His way of working was "ecological," emphasizing predation and its influence on protective adaptations. The food-web of wild animals was Wallace's point of departure in contemplating the evolution of colors and their uses.

Wallace and Darwin were also considerably different in personality, as well as in their positions on scientific questions in evolutionary theory. The differences were obvious enough to cause several writers to distinguish "Wallaceism" from "Darwinism." It could not fail to cause comment that Wallace denied natural selection's adequacy to cause the evolution of the most important animal of all--*Homo sapiens*, or that Darwin included Lamarckian explanation and an "aesthetic" component in his theory of sexual selection.

The differences between Wallace and Darwin have been discussed less often than their similarities in historical studies. The nineteenth-century writers who were closer to the central characters in the Darwinian drama tended to be struck by their differences and with good cause. The contrast between the two Victorian naturalists is clearly portrayed in the way each man concluded his work on the selection of colors and sexual dimorphism. Darwin ended the *Descent of Man* with the sentence:

> We must, however, acknowledge, as it seems to me, that man with all his noble qualities, with sympathy which feels for the most debased, with benevolence which extends not only to other men but to the humblest living creature, with his god-like intellect which has penetrated into the movements and constitution of the solar system--with all these exalted powers--Man still bears in his bodily frame the indelible stamp of his lowly origin.[122]

Wallace concluded the chapter on plant colors in *Darwinism* with Alfred Tennyson's poem "Flower in the Crannied Wall."[123]

His essay on animal colors was also brought to an end with a poem. It was like him to emphasize the "marvel and mystery" of nature study in his last major statement on the connection of Darwinism to particular data. He wrote:

> The marvel will ever remain to the sympathetic student of Nature; but I venture to hope that in the preceding chapters I have succeeded in lifting--if only by one of the corners--the veil of mystery which has for long shrouded this department of Nature.[124]

<u>On a Peacock's Feather</u>.[125]

In Nature's workshop but a shaving,
   Of her poem but a word,
But a tint brushed from her pallette,
   This feather of a bird!
Yet set it in the sun-glance,
   Display it in the shine,
Take graver's lens, explore it,
   Note filament and line,
Mark amethyst to sapphire,
   And sapphire to gold,
And gold to emerald changing
   The archetype unfold!
Tone, tint, thread, tissue, texture,
   Through every atom scan,
Conforming still, developing,
   Obedient to plan.
This but to form a pattern
   On the garment of a bird!
What then must be the poem,
   This but its lightest word!
Sit before it, ponder o'er it,
   'Twill thy mind advantage more
Than a treatise, than a sermon,
   Than a library of lore.

Notes--Chapter V

ANOTHER DAY'S LIFE:

DARWIN AND WALLACE ON SELECTION

1. George John Romanes (1848-1894) physiologist, comparative psychologist, evolutionary theorist. See *Dictionary of Scientific Biography* article by John Lesch.

2. G. J. Romanes, "Mr. Wallace on Darwinism," *Contemporary Review* 1889:244-258; p. 248.

3. G. J. Romanes, *Darwin and After Darwin. An Exposition of the Darwinian Theory and a Discussion of Post-Darwinian Questions*, 3 vols. I, 1892; II, 1895; III, 1897. (Chicago: Open Court)

4. Ambrosius Arnold Willem Hubrecht (1853-1915) zoologist, comparative embryologist, "Darwinism *versus* Wallaceism," *Contemporary Review* 94:629-634, 1905.

5. Hugo De Vries (1848-1935) plant hybridist, a rediscoverer of Gregor Mendel's work on heredity. Author of *Intracellulare Pangenesis* (Jena, 1889), and *Die Mutationstheorie. Versuche und Beobachtungen uber die Enstehung von Arten im Pflanzenreich* (Leipzig, 1903).

6. Hubrecht, "Darwinism *versus* Wallaceism," p. 630.

7. Alfred Russel Wallace, "Darwinism *versus* Wallaceism," *Contemporary Review* 94:716-717, 1908.

8. Wallace tried to explain to Darwin that he adopted a new position on human evolution because new "phenomena" necessitated the revision of the theory. The phenomena were psychical and spiritualistic phenomena. Wallace wrote to Darwin, April 18, 1869: "My opinions on the subject [of human evolution] have been modified solely by the consideration of a series of remarkable phenomena, physical and mental, which I have now had every opportunity of fully testing, and which demonstrate the existence of forces and influences not yet recognized by science." James Marchant, *Alfred Russel Wallace: Letters and Reminiscences* (London: Cassel & Co., 1916), p. 200. (Hereinafter referred to as "Marchant.")

9. Letter from Darwin to Wallace, February 26, 1887. "The analogy of birds makes me believe in sexual selection

with respect to Insects." Marchant, p. 148.

10. Charles Darwin, The Descent of Man and Selection in Relation to Sex (London: John Murray, 1871). The chapters on birds are 13, 14, 15, 16. (Hereinafter referred to as Descent.)

11. Charles Darwin and Alfred Wallace, "On the tendency of Species to form Varieties; and On the Perpetuation of Varieties by Natural Means of Selection," Journal Linnean Society of London 3:45-62, 1858. Quotation from p. 50. "Extract from an unpublished work on Species by C. Darwin, consisting of a portion of a chapter entitled 'On the Variation of Organic Beings in a State of Nature; on the Natural Means of Selection; on the Comparison of Domestic Races and True Species."

12. Charles Darwin, On the Origin of Species, 1st ed., p. 89.

13. Charles Darwin, The Autobiography of Charles Darwin, 1809-1882, ed. by Nora Barlow (New York: W. W. Norton, 1969), p. 131. (Hereinafter referred to as Autobiography.)

14. Charles Lyell, The Geological Evidences of the Antiquity of Man, with Remarks on Theories of the Origin of Species by Variation (London: John Murray, 1863); and Thomas Henry Huxley, Evidence as to Man's Place in Nature (London, 1863).

15. Darwin, Autobiography, p. 131.

16. Letter, Darwin to Wallace, October 21, 1869. Marchant, p. 203.

17. Letter, Darwin to Wallace, March 7, 1860. Marchant, p. 116.

18. Alfred R. Wallace, "The Origin of Human Races and the Antiquity of Man Deduced from the Theory of Natural Selection," Anthropological Review 2:clvii-clxx (May 1864).

19. Letter, Darwin to Wallace, May 28, 1864. Marchant, p. 128.

20. Alfred Russel Wallace, The Malay Archipelago: the Land of the Orang-utan, and the Bird of Paradise. A Narrative of Travel, with Studies of Man and Nature (London: Macmillan, 1869).

21. Letter, Wallace to Darwin, May 29, 1864. Marchant, p. 130.

22. Letter, Darwin to Wallace, June 15, 1864. Marchant, p. 132.

23. Notes offered on May 28, 1864, and withdrawn June 15, 1864.

24. The letter from Darwin to Wallace, June 22, 1866 (Marchant, p. 138) contains these remarks from Darwin: "I cannot conceive that the most firm believer in Species could read it without being staggered. Such papers will make many more converts among naturalists than long-winded books such as I shall write if I have strength...I have been particularly struck with your remarks on dimorphism...How can one female form be selected and the intermediate forms die out, without also the other extreme form also dying out from not having the advantages of the first selected form? for I understand, both female forms occur on the same island." It is a little surprising that Wallace was not more concerned with how adaptive variations could be preserved in a species without replacing totally the parental form. Divergence in a single species is perhaps more difficult to explain on the laws of heredity than between species. In his answer to Darwin February 4, 1866,(Marchant, p. 139), Wallace replied that both the typical and variant forms have protective adaptations and are both preserved by natural selection. For Wallace, protection determines which variations will survive in the species; the real quandry is "genetic." He wrote: "The physiological difficulty is to me greater, of how each of two forms of female produces offspring like the other female as well as like itself, but no intermediates."

25. Letter, Darwin to Wallace, February 23, 1867. Marchant, pp. 146-147. Including the question: "If anyone objected to male butterflies having been made beautiful by sexual selection, and asked why should they not have been made beautiful as well as their caterpillars, what would you answer? I could not answer, but should stand my ground."

26. Wallace's letter is missing, but Marchant quoted from _My Life_ II, p. 3, where Wallace recounted the incident. Wallace was just then preparing the paper on mimicry for the _Westminster Review_ and had been thinking about conspicuous coloration in butterflies and moths as signals of their being unpalatable to predators. He could, therefore, answer Darwin almost at once (Marchant, pp. 147-148)

27. Letter, Darwin to Wallace, February 26, 1867. Marchant, p. 148.

28. Peter Vorzimmer, _Charles Darwin: The Years of Controversy. The Origin of Species and its Critics, 1859-1882_ (Philadelphia: Temple University Press, 1970), p. 193.

29. Marchant, p. 151-152.

30. Marchant, p. 152.

31. Letter, Wallace to Darwin, May 1, 1867. Vorzimmer, _Charles Darwin: The Years of Controversy_, p. 195.

32. Letter, Darwin to Wallace, May 5, 1867. Marchant, pp. 152-154.

33. Marchant, p. 154.

34. "_Naturalist_" _Philosopher_ might have been even better.

35. Letter, Darwin to Wallace, April 15, 1868. Marchant, p. 175.

36. Letter, Darwin to Wallace, April 30, 1868. Marchant, p. 177.

37. Letter, Darwin to Wallace, September 16, 1868. Marchant, p. 183.

38. Letter, Wallace to Darwin, September 27, 1868. Marchant, p. 188.

39. Wallace seems to have answered Darwin's letter of the 23rd of September, 1868, twice. The first reply was dated September 27, 1868, and the second reply October 4, 1868. Darwin's letter shows that he is worried about genetic problems. Darwin wrote: "We differ, I think, chiefly from fixing our minds perhaps too closely on different points, on which we agree: I lay great stress on what I know takes place under domestication: I think we start with different fundamental notions on inheritance." (Marchant, p. 185.) Darwin was trying to sort out two different genetic situations which both have sexual dimorphism as their result. First, variation that appears initially in one sex only and is transmitted to offspring of the same sex; and second, variations which arise in one sex, usually the male, and are transmitted to both sexes of offspring but are _latent_ in their

expression in offspring of the opposite sex unless revealed by diseases or deterioration of the ovaries with age.

40. Letter, Wallace to Darwin, March 11, 1871. Marchant, pp. 212-213. Wallace wrote in My Life, II, p. 14: "It is quite pathetic how much he [Darwin] felt difference of opinion from his friends. I, of course, should have liked to have been able to convert him to my views, but I did not feel it so much as he seemed to do."

41. The Academy has been classified as a monthly, scientific periodical, religiously neutral and aimed at a highly educated readership. It had a circulation of some 8,000 according to Alvar Ellegard. Darwin and the General Reader (Gotëborg: Elanders, 1958), p. 369.

42. Letter, Wallace to Darwin, March 11, 1871. Marchant, pp. 212-213.

43. A. R. Wallace, [Review of] "The Descent of Man and Selection in relation to Sex," Academy 2:177-183, 1871.

44. Wallace, My Life, II, p. 17.

45. Grant Allen, The Colour Sense: Its Origin and Development. An Essay in Comparative Psychology (London: Trubner, 1879).

46. Allen, The Colour Sense, p. 281.

47. Allen, The Colour Sense, p. 282.

48. Even though Romanes was much less anthropomorphic than Allen, his work was criticized on just that point by his successors. See C. Lloyd Morgan, Animal Life and Intelligence (London: Edward Arnold, 1890-91).

49. Wilma George, Biologist Philosopher. A Study of the Life and Writings of Alfred Russel Wallace (London: Abelard-Schuman, 1964), pp. 210-211.

50. Alfred Russel Wallace, "Colour in Nature," Nature 18:501-504, 1879.

51. Wallace, My Life, II, p. 22.

52. Alfred Russel Wallace, Contributions to the Theory of Natural Selection. A Series of Essays.(London: Macmillan, 1870). The edition used here is the 2nd edition of 1871

(hereinafter referred to as Contributions).

53. Wallace, Contributions, p. 203.

54. Wallace, Contributions, Chapter VIII, "Creation by Law" was first published as an article in the Quarterly Journal of Science 1867. The Duke of Argyll [George Douglas Campbell] (1823-1900), The Reign of Law (London: George Routledge, 1867). The fifth edition used here is of 1869.

55. Duke of Argyll, Reign of Law, p. 248.

56. Wallace, Contributions, p. 283.

57. Alfred Russel Wallace, Darwinism. An Exposition of the Theory of Natural Selection with Some of Its Applications (London: Macmillan, 1889). Edition used here is 1891 and hereinafter referred to as Darwinism. Quotation from p. 154.

58. "The Philosophy of Birds' Nests," first published in Intellectual Observer 11:413-420, 1867; and, "A Theory of Birds' Nests: Showing the Relation of Certain Sexual Differences of Colour in Birds to their Mode of Nidification," first published in Journal of Travel and Natural History I: 73-89, 1868-69.

59. Wallace, "A Theory of Birds' Nests," J. of Travel, p. 78.

60. Wallace, "A Theory of Birds' Nests," J. of Travel, pp. 83-84.

61. Wallace, "A Theory of Birds' Nests," J. of Travel, p. 82.

62. Wallace, "A Theory of Birds' Nests," J. of Travel, p. 82.

63. Charles Darwin, The Variation of Animals and Plants under Domestication. Originally published by John Murray, 1878. The edition used here is (New York: D. Appleton, 1896). See Volume II, Chapter XXVII, "Provisional Hypothesis of Pangenesis."

64. The similarity of the adult female to the young of both sexes, Darwin argued, supported the conclusion that the male had departed further from the ancestral form. Young animals and embryos were understood to be closer to the ancestral type.

65. Wallace, _Contributions_, Chapter 3, "Mimicry, and Other Protective Resemblances Among Animals," pp. 81-82.

66. Andrew Murray, "Reply to Mr. Wallace's Theory of Birds' Nests," _Journal of Travel and Natural History I_: 137-45, 1868-69; and, The Duke of Argyll, "On Mr. Wallace's Theory of Birds' Nests," _Journal of Travel and Natural History I_:277-287, 1868-69.

67. Argyll, "On Mr. Wallace's Theory," _J. Travel_, p. 287.

68. Argyll, "On Mr. Wallace's Theory," _J. Travel_, p. 283.

69. Argyll, "On Mr. Wallace's Theory," _J. Travel_, p. 286.

70. Murray, "Reply to Mr. Wallace's Theory," _J. Travel_, p. 140.

71. Murray, "Reply to Mr. Wallace's Theory," _J. Travel_, p. 143.

72. Murray, "Reply to Mr. Wallace's Theory," _J. Travel_, p. 144. The allusion is to Psalm 139:14, King James Version. "I will praise thee; for I am fearfully and wonderfully made: marvelous are thy works; and that my soul knoweth right well."

73. Andrew Murray, "On the Disguises of Nature: being an Inquiry into the Laws which regulate External Form and Colour in Plants and Animals," _Edinburgh New Philosophical Journal 11_:66-90, 1860.

74. For a brief discussion of the views of biologists in the 1960's, see Wilma George, _Biologist Philosopher_, pp. 214-215.

75. Darwin, _Origin of Species_, 6th edition. Chapter VIII, "Instinct," first page of chapter.

76. Charles Darwin, _Descent of Man_, 2nd edition of 1874, p. 478 (page numbers for the second edition are from Modern Library edition). On the inheritance of moral sensitivity in people, Darwin summarized: "Looking to future generations, there is no cause to fear the the social instincts will grow weaker, and we may expect that virtuous habits will grow stronger, becoming perhaps fixed by inheritance" (p. 494). On p. 446, men and animals are compared. "As man possesses the same senses as the lower animals, his fundamental intuitions must be the same. Man has also some few instincts in common, as that of self-preservation, sexual love, the love

of the mother for her new-born offspring, the desire possessed by the latter to suck, and so forth. But man, perhaps, has somewhat fewer instincts than those possessed by the animals which come next to him in the series."

77. Michael T. Ghiselin, The Triumph of the Darwinian Method (Berkeley: University of California Press, 1969), p. 202.

78. Julian Huxley, "The Emergence of Darwinism," Journal of the Linnean Society (Zoolcgy) 44:1-13, 1958. Quotation on p. 11.

79. Charles Darwin, "Sexual Selection in Relation to Monkeys," Nature 15:18-19, 1876; and, "The Sexual Colors of Certain Butterflies," Nature 21:237, 1880.

80. W. Van Dyck, "On the Modification of a Race of Syrian Street-Dogs by Means of Sexual Selection," with a Preliminary Notice by Charles Darwin," Proceedings of the Zoological Society 25:367-370, 1882.

81. Alfred Tylor (1824-1884), other publication in geology. Colouration in Animals and Plants (London: Alabastar, Passmore, & Sons, 1886).

82. See Chapter I above. Rev. Henry Hugh Higgins (1814-1893), "On Colour Patterns in Natural Productions," Proceedings of the Liverpool Literary and Philosophical Society 1857, pp. 132-143.

83. Tylor claims to have been influenced by Samuel Butler's Life and Habit (London: Trubner, 1878).

84. Tylor, Colouration in Animals and Plants, p. 15.

85. Tylor, Colouration in Animals and Plants, p. 97.

86. Tylor, Colouration in Animals and Plants, p. 18.

87. Tylor, Colouration in Animals and Plants, p. 81. Orthodox Darwinians can only have been appalled. Tylor wrote: "If we grant [inherited] memory we eliminate this hap-hazard natural selection. We see how a being that has once begun to perform a certain action will perform it automatically, and when its habits are confirmed its descendants will more readily work in this direction than any other, and so specialization may arise."

88. A. R. Wallace, "The Colors of Animals and Plants," The American Naturalist 11:641-662, 713-727. Also in Macmillans Magazine, pp. 364-408, 464-471,(September 1877).

89. Wallace, "The Colors of Animals and Plants," American Naturalist 11, p. 650.

90. Wallace, "The Colors of Animals and Plants," American Naturalist 11, p. 714. The sentence is preserved unchanged in the chapter of Tropical Nature, "The Colours of Animals and Sexual Selection," Tropical Nature and Other Essays (London: Macmillan, 1878). Page numbers here are to the second edition of Natural Selection and Tropical Nature Essays on Descriptive and Theoretical Biology (London: Macmillan, 1891). Quotation from page 366.

91. Wallace, "The Colors of Animals and Plants,"" American Naturalist 11, pp. 715-716. Darwin, Descent of Man, 2nd ed., pp. 853-854.

92. Wallace, "The Colors of Animals and Plants," American Naturalist 11, p. 719.

93. Darwin, Descent of Man, 2nd ed., p. 658.

94. Wallace, "The Colours of Animals and Plants," American Naturalist 11, p. 725.

95. Wallace, Natural Selection and Tropical Nature, 1891, p. 349.

96. Wallace, Darwinism. An Exposition of the Theory of Natural Selection with Some of Its Applications (London: Macmillan, 1889). Edition used here is that of 1891, p. iii.

97. Gerald M. Henderson, "Alfred Russel Wallace: His Role and Influence in Nineteenth Century Evolutionary Thought" (unpublished Ph.D. dissertation, University of Pennsylvania, 1960), pp. 66-67.

98. Wallace, My Life, II, p. 201.

99. Wallace, Darwinism, p. iv.

100. Wallace, Darwinism, p. 6. "We claim for Darwin that he is the Newton of natural history, and that, just so surely as that the discovery and demonstration by Newton of the law of gravitation established order in place of chaos and laid a sure foundation for all future study of the starry

heavens, so surely has Darwin, by his discovery of the law of natural selection and his demonstration of the great principle of the preservation of useful variations in the struggle for life, not only thrown a flood of light on the process of development of the whole organic world, but also established a firm foundation for all future study of Nature."

101. Wallace, *Darwinism*, pp. 204-205.

102. Wallace, *Darwinism*, p. 174.

103. Edward Bagnall Poulton (1856-1943), entomologist, winner of the Royal Society's Darwin Medal in 1914. Author of *Essays on Evolution, 1889-1907* (Oxford: Clarendon Press, 1908), and many other works on evolution both historical and experimental. One obituary biography is by G. D. Hale Carpenter, in *Proceedings of the Linnean Society of London*, *156*:219-223, 1943-44.

104. Edward Bagnall Poulton, *The Colours of Animals. Their Meaning and Use, Especially Considered in the Case of Insects* (New York: D. Appleton, 1890).

105. Poulton, *Colours of Animals*, p. 336. G. D. H. Carpenter has suggested that the Greek names for the uses of color would have the good effect of avoiding any hint of volition in the term "mimicry"; that seems like the probable reason Poulton had for selecting his words which had no connotations.

106. Poulton, *Colours of Animals*, p. 338. This table is after Poulton's.

107. Hugh B. Cott, *Adaptive Coloration in Animals* (London: Methuen, 1940).

108. Poulton, *Colours of Animals*, p. 308.

109. Julian Huxley, "Darwin's Theory of Sexual Selection and the Data Subsumed by it, in the Light of Recent Research," *American Naturalist* *72*:416-433, 1938. The most complete recent work is *Sexual Selection and the Descent of Man, 1871-1971*, edited by Bernard Campbell (Chicago: Aldine Publishing Co., 1972).

110. August Weismann, *Studies in the Theory of Descent*. Translated and Edited, with Notes, by Raphael Meldola. With a Prefatory Notice by Charles Darwin (London: Sampson, Low, Marston, Searle, & Rivington, 1882).

111. August Weismann, *The Evolution Theory*. Translated by J. Arthur Thomson and Margaret R. Thomson (London: Edward Arnold, 1904). Volume I (of two volumes) contains the discussions of animal and plant color.

112. William Bateson, *Materials for the Study of Variation Treated with Especial Regard to Discontinuity in the Origin of Species* (London: Macmillan, 1894).

113. Gustav Heinrich Eimer, *Organic Evolution as the Result of the Inheritance of Acquired Characters According to the Law of Organic Growth*. Translated by J. T. Cunningham (London: Macmillan, 1890); and, Alpheus S. Packard, "The Origin of the Markings of Organism (Poecilogenesis) Due to the Physical Rather than to the Biological Environment with Criticisms of the Bates-Müller Hypothesis," *Proceedings of the American Philosophical Society* 43:393-450, 1904.

114. Marion I. Newbigin, *Colour in Nature. A Study in Biology* (London: J. Murray, 1898).

115. Frank Beddard, *Animal Colouration, an Account of the Principle Facts and Theories Relating to the Colours and Markings of Animals* (London: Swan Sonnenschein, 1892).

116. Beddard, *Animal Colouration*, p. 92.

117. Letter, Wallace to E. B. Poulton, August 5, 1904. Marchant, p. 328.

118. One example from the work of each individual relating some aspect of animal colors and their uses to the synthetic theory of evolution is given here. G. D. H. Carpenter, *Mimicry* (London: Methuen, 1933). E. B. Ford, "The Genetic Basis of Adaptation," in *Evolution*, edited by Gavin DeBeer (Oxford: Clarendon Press, 1938), pp. 48-55. E. B. Poulton, "Insect Adaptation as Evidence of Evolution by Natural Selection," in *Evolution*, ed. by Gavin DeBeer, pp. 1-10. E. Mayr, *Systematics and the Origin of Species*, see Chapter 4 [geographic variation and the genetics of polymorphism] (New York: Columbia University Press, 1942). J. Huxley, "The Present Standing of the Theory of Sexual Selection," in *Evolution*, ed. by Gavin DeBeer, pp. 11-42. R. A. Fisher, *The Genetical Theory of Natural Selection*, see Chapter 6 [Sexual Selection] and 7 [Mimicry] (Oxford: Clarendon Press, 1930).

119. J. Huxley, "The Present Standing of the Theory of Sexual Selection," in *Evolution*, ed. by Gavin DeBeer, pp.

1-10. Later authors have advanced the theory considerably; see volume edited by Bernard Campbell, Sexual Selection and the Descent of Man, 1871-1971.

120. David Lack, Life of the Robin (London: H. F. & G. Nitherby, 1944); and Konrad Lorenz, On Agression (New York: Bantam Books, 1966). See Wilma George, Biologist Philosopher, pp. 212-218, for an assessment of work on bird behavior and aggressive display. She notes that threat display is only a special case of recognition as Wallace knew it was. She mentions several times that Wallace's contributions to the subject are almost totally overlooked in modern works on the subject. She wrote: "The present attitude to colour should be attributed to both Darwin and Wallace, therefore, even though Wallace's name has dropped out of the controversy. Recognition colours seem to be of far greater importance than most nineteenth-century naturalists believed and, although many biologists would find it difficult to believe for the lack of reference to him, the first realization of the importance of recognition colours is due entirely to A. R. Wallace." Biologist Philosopher, p. 218. Also see, A. J. Meyerriecks, Man and Birds: Evolution and Behavior (New York: Bobbs-Merrill, 1972).

121. H. B. D. Kettlewell, "Darwin's Missing Evidence," Scientific American 200:48-53, 1959.

122. Darwin, Descent of Man, 2nd ed., p. 920.

123. Wallace, Darwinism, p. 228.

Flower in the crannied wall,
I pluck you out of the crannies;
I hold you here, root and all, in my hand,
Little flower; but if I could understand
What you are, root and all, and all in all,
I should know what God and man is.

124. Wallace, Darwinism, p. 202.

125. Wallace, Darwinism, p. 203. Wallace did not identify the poet except as a "poet-artist."

CONCLUSION

Julian Huxley reduced the Darwinian theory of evolution to its bare essentials when he wrote in Evolution: The Modern Synthesis that Darwin's theory is based on three observable facts from nature and on two deductions from these facts.[1] Darwin's first fact was that animals naturally tend to increase their numbers according to a geometrical ratio; the second fact was that most species actually have a rather stable population despite the production of offspring according to the geometrical ratio. The conclusion drawn from these observations is that there is a struggle for life (and for reproduction) resulting from competition between individuals which are produced in greater numbers than can survive. Darwin's third observation was that individual organisms vary, sometimes quite noticeably. The second deduction was based on the struggle for existence (the first deduction) and on variation (the third fact). Darwin deduced that certain variations are more favorable to survival than others, allowing those individuals possessing them to survive and reproduce. This differential survival was called "Natural Selection" by Darwin and Wallace. Natural selection, then, is the name given to the maintenance and proliferation of variations which equip organisms to cope with their environments.

The adaptive coloration of animals is related

specifically to Darwin's third observation, variation, and to his second deduction, natural selection. The relationship between variations in individuals and natural selection was taught to the first generation of naturalists after the Origin of Species by using examples of differential survival of individuals with favorable variations in color. The concept of natural selection is still taught to students by instructors using very similar data to exemplify the concept.

According to Darwinians, the simplest color phenomena, like the white Arctic animals and the sandy-colored desert species, indicated that differential survival based on variation provided an adequate basis for gradual change in species. The Darwinian theory, however, provided no real clue to the origin of variation. In the middle of the nineteenth century there were already other theories on the origin of variations and of species. The Darwinian theory was not at all launched into a void fulfilling a widely felt need for explaining the unexplained. In England, perhaps more than in any other European country, natural theology was, in fact, thriving into the 1840's. Naturalists who were disposed toward exclusively naturalistic explanation had the alternative of accepting Lamarckian explanation of adaptation and of organic change.[2]

The general matching of white animals to snowy environments, for example, had been long explained as indicating Providential forethought on the part of the creator.

The work of William Kirby and William Spence in their Introduction to Entomology gave the most complete statement of natural theological perspective on entomological observations. As observers, natural theologians were second to none in their accuracy. Their empirical concern as scientists was supported by their Christian values. The natural theologians discussed quite thoroughly many of the phenomena later associated with the Darwinian controversy.

Natural theologians were aware that nature was complex; that it presented far greater challenges than the mere explanation of general camouflage in animals which shield them from excessive predation. Kirby and Spence discussed mimetic resemblances in which likenesses occurred between different orders of insects. Westwood also described the mimicry between distinct families of butterflies in a work on the systematics of Lepidoptera.[3] The taxonomic concerns of the early nineteenth century were one way of expressing a very real interest in the organization of life and in the discovery of its meanings. Mimicry was first part of taxonomy and only after Bates's paper of 1862 did it become part of the Darwinian controversy. As Darwinians gained adherents and as experiments were initiated to test evolution and natural selection, taxonomy and natural theology declined together.

Without natural theology, taxonomy was understood by young zoologists to be nothing but "dry bones."[4] Charles

Gillispie expressed that prejudice when he wrote, "Taxonomy little tempts the historian of scientific ideas. The problems were fussy and practical, but the question whether classifications are natural or artificial did not ultimately prove interesting."[5] The statement shows a profound misunderstanding of both taxonomy as a study and of the nineteenth-century naturalists who classified organisms.

The entire century was marked by "fussiness" and by "practicality;" these qualities were then sources of pride, not of shame. The careful taxonomic work required a very careful plodding approach. It is important not to miss the fact that taxonomy was not always done in isolation from broader questions. Not only was taxonomy seen as a *sine qua non* to other zoological studies, it was intimately connected to the developing understanding of the Creator and His involvement in the natural economy. Taxonomy made the hierarchy of nature concrete and witnessed to the existence and nature of God.

Nevertheless, the questions asked and answered by natural-theologically inclined taxonomists were not very different from those asked and answered by the first evolutionists after Darwin. The taxonomist who works toward the placing of species in proper relations in a created hierarch based on structural likenesses and differences works with similar characters as those used by the evolutionary zoologist. Darwin did not find the problems of taxonomy sterile and the discipline moribund when he produced his prodigious volumes on the Cirripedia.[6] Naturalists like Darwin, Kirby, and Westwood realized, as Charles Gillispie does not, that a proper taxonomy epitomizes and demonstrates a view of the organism as a whole including both functional and morphological characteristics.

The inverse relation of the study of taxonomy and the rise of interest in evolutionary theory and experimental technique is very interesting. The fruitful questions asked by taxonomy in the period just before Darwin were posed in quite a different context in the years of Darwinian controversy. It is by studying a body of connected data that one can see the intimate connection between observations, i.e., "facts," and theories. Theories, in turn, have had important connections to the development of disciplines--"schools" of thought, and changes in techniques and methods of attacking questions. If taxonomy gave its leading place in zoological studies to evolutionary "comparative" studies, "comparative" work yielded its place with the rise of genetics. Although early geneticists were usually evolutionists, they were rarely orthodox Darwinians. In fact, they saw large mutations as the major or exclusive mechanism of evolutionary change. Natural selection was often given only a very minor role in geneticists's works on evolution if it was given any role at all.

The advocates of differing theories all used the

phenomena of adaptive coloration in animals in order to argue that their theory was better than other competing theories in accounting for the evolution of such diverse phenomena. Henry Walter Bates and A. R. Wallace were among the earlier naturalists to give accounts of adaptive coloration of insects without appealing to the argument from design. Darwin's efforts in the Origin of Species to present the scope of evolution was supplemented by the studies of particular species and families of organisms.

Field zoologists who were involved in studies of the geographical distribution of animals were very prominent in the effort to interpret the data of variation in wild populations. In the early years of the Darwinian controversy the work on geographical variation provided some of the most compelling evidence against the concept of permanent, fixed, created species. It was, also, the collectors and field naturalists who could understand the struggle for existence from their experiences of observing the chains of predation in natural systems. Struggle and differential survival were explained to scientific readers by Wallace and Bates in their studies of mimicry and polymorphism in two families of butterflies, the Papilionidae of the Malay Archipelago and the Heliconiidae of the Amazonian region of South America. The general question of the origin of species was treated with great care in a few organisms in the hope of throwing light on the general question of the evolution of populations diverging from ancestral stocks. Bates demonstrated as clearly as he could that geographical series refuted the conception of species as permanently created units and that the mimicry of protected butterflies by unprotected species had come about by predation rather than by design. Wallace also combined geographical studies with assessments of the species concept and moved on to make the claim that phenomena of animal colors give witness to the history of the organisms and its long course of natural selection of useful variations.

The history of the mimicry theory is a microcosmic way of understanding the history of the acceptance of the Darwinian theory of evolution by natural selection. Mimicry was used by its Darwinian proponents to teach other naturalists the concepts it incorporates, and it was one of the earliest major issues in the reception of evolutionary theory. The discoverers of mimetic phenomena almost invariably described their data in terms of natural secondary causes. Mimicry was a subject that became associated with the naturalistic modes of explanation used in Darwinian evolutionary theory. Work like Bates's on the geographical variation of mimetic butterflies confronted the species question with an overt denial of the traditional belief in natural history that species were created in the form in which they are presently seen. The students of mimicry connected phenomena of variation with differential survival--natural selection.

Despite every effort by a growing contingent of Darwinian evolutionists, the distinction between evolution in general and its mechanism, natural selection, was continually confused and obfuscated throughout the nineteenth century and well into the twentieth century. Much of the so-called "Darwinism" in the period is actually evolution without its characteristic Darwinian component, natural selection. In meeting after meeting of the Entomological Society of London in the 1860's and 1870's Bates and Wallace used examples from papers read and from displays of insects on exhibit to illustrate how natural selection can result in species change. Their explanations were greeted coldly by those who refused any theory which did not explain the origin of the variations it required and was based on "change."

In the years following the <u>Origin of Species</u>, the use in science of the argument from design declined sharply, partly through Darwin's influence. Naturalistic explanations of biological phenomena became widely acknowledged to be a desideratum for scientific work. But even when naturalistic modes of explanation were used, it was not invariably the case that Darwinian natural selection would be the theory of choice. Neo-Lamarckianism, in vogue in America, and mutation theory vied with Darwinism for acceptance by biologists. Many elements in science and society seemed to conspire to slow the pace at which natural selection was understood and accepted by the scientific community. Some of the reasons were internal to science, but not all.

Devotion to inductive science in England was something of a national crusade. The Victorian scientists of the 1860's and 1870's were devoted to the advancement of descriptive, observational, "Baconian Science." George Boas was correct to caution that "The nineteenth century did not get heated up over evolution because it was the nineteenth century or the Victorian period."[7] The challenges to traditional assumptions in the period were profound. The scientists in the period were confronted in Darwinian theory with a naturalistic theory which seemed to many to exclude design, and to cast doubt on the existence of a provident Designer. It is not surprising, therefore, that many people believed there would be deleterious results for society if such a theory were approved by scientists. The design of animals harmonizing with their environments had long been heralded as proving that the world was made according to a plan. The challenge of Batesian mimicry had echoes reaching back to very basic philosophical premises. If butterflies' color patterns were not designed; if species are not created; then what is designed or created? The people who got "heated up over evolution" in the nineteenth century were well justified. Whole world views are challenged by particular data. In his work, "The Influence of Darwin on Philosophy," John Dewey described the decline of concern about both alternatives in the debate about design and chance:

> Old ideas give way slowly...The conviction persists--though history shows it to be a hallucination-- that all the questions can be answered in terms of the alternatives that the questions themselves present. But in fact intellectual progress usually occurs through sheer abandomnent of questions together with both of the alternatives they assume...We do not solve them: we get over them. Old questions are solved by disappearing, evaporating, while new questions corresponding to the changed attitude of endeavor and preference take their place. Doubtless the greatest precipitant of new methods, new intentions, new problems, is the one effected by the scientific revolution that found its climax in the "Origin of Species."[8]

The _Origin of Species_ certainly represents a major watershed in the history and philosophy of science. Fewer questions were dissolved by the Darwinian revolution than Dewey may have realized, however. Sometimes the context in which old ideas are discussed becomes so altered that questions are no longer recognized for what they are: old questions. The controversies involved in the evolutionary theory--of design _versus_ chance, natural explanation _versus_ supernatural, empiricism _versus_ idealism--were too grand in their proportions to be solved by any single effort, however heroic it might be. The revolution Darwin had begun with the _Origin of Species_ was only slowly resolved into a return to the doing of normal science in which natural selection was an accepted prominent principle.

Darwin's role in the discussion of evolution and

adaptive coloration is rather complex. He had a great interest in the extension of natural selection, yet in the discussions of animal colors he generally emphasized its inadequacy to explain the development of sexually dimorphic traits. Darwin's views on sexual selection depending on female choice became the subject of a long controversy with Wallace. The evolution of colors, patterns, and decorative appendages in animals was appraised very differently by the two chief "Darwinians." The disagreement between the two was largely a product of each man's understanding of the relations of wild and domestic species. Contrasting views of inheritance, varying uses of the principle of utility, and important differences in the animals studied as models forming the basis for the generalized theory, gave each theorist's work its distinctive character. Darwin's theory of sexual selection depended heavily on his bird studies; whereas Wallace's most careful work was with Malayan butterflies and to a lesser extent with wild birds. Wallace eventually discounted almost completely the relevance of domestic species for understanding wild animals. The colors of animals was a useful subject in Darwin's and Wallace's private repartee about selection as well as in the public forum of published papers.

The data of animal coloration, like any other class of data, is "theory-laden." The perspective of the observer is important in determining the "facts" themselves. For example, Bates and Westwood both knew about mimicry as analogous resemblances, but their perceptions were altered by their presumptions. Theoretical and philosophical frameworks change radically the meaning of observations. The history of theories about animal color provides a good example to illustrate the changing relations of data to theory. The Victorian scientists who first grappled with the problem of reconciling or refuting the connection of adaptive colors of animals to the Darwinian theory were dealing with a major issue common to science and philosophy. Percy Bridgman wrote: "We do not have a simple event A causally connected with simple event B, but the whole background of the system in which the events occur is included in the concept, and is a vital part of it."[9] The complexity of the relationships between facts and interpretations which Bridgman emphasized is found in the study of the history of evolutionary theory. The publication of the _Origin of Species_ in November of 1859 was not a simple event follwed by other simple events, like Westwood's publication of a negative review in the _Gardener's Chronicle_, for example.[10] The "background of the system" of which Bridgman wrote plays too profound a part in the events of the history of the evolutionary debates to be reduced to a simple series of disconnected, discrete events. The accepted approach in the natural history disciplines of the nineteenth century was to approach the great problems through the solution of smaller, more accessible problems. For the true Baconian of the nineteenth century each fact and question bears on the eventual understanding of the whole. Although opposite in their metaphysical assumptions, both Charles Darwin and John Obadiah Westwood were profoundly concerned with the resolution of the problem of mimetic analogies. The question of mimicry for them clearly counted for or against the idea of permanently created species and God's role in nature. The nineteenth-century debate about the meaning of animal coloration was a controversy about particular data, but it had very far-reaching meaning for the debaters.

Notes

CONCLUSION

1. Julian Huxley, *Evolution: The Modern Synthesis* (New York: John Wiley, 1964), p. 14. First published in 1942.

2. Herbert Spencer is one who preferred Lamarckian explanation to natural theology before the *Origin of Species*.

3. Edward Doubleday and John Obadiah Westwood, *The Genera of Diurnal Lepidoptera*, 2 vols. (London: Longman, Brown, Green & Longmans, 1846-1842).

4. G. D. Hale Carpenter used the term in his obituary of Poulton in stating what Poulton disliked in the style of entomology done at Oxford by Westwood. *Proceedings of the Linnean Society of London* 156: 220, 1943-44.

5. Charles C. Gillispie, *The Edge of Objectivity. An Essay in the History of Scientific Ideas* (Princeton, New Jersey: Princeton University Press, 1960), p.170.

6. Charles Darwin, *A Monograph of the Sub-Class Cirripedia, with Figures of all Species. The Lepadidae; or, Pedunculated Cirripedia* (London: the Ray Society, 1851); and *The Balanidae, or Sessile Cirripedes; The Verrucidae, etc.* (London; Ray Society, 1854).

7. George Boas, *The History of Ideas* (New York: Charles Scribner's Sons, 1969, p.141.

8. John Dewey, "The Influence of Darwin on Philosophy," in *Contemporary Thought* (New York, 1910). Quoted from the excerpt in *Darwin: A Norton Critical Edition*, selected and edited by Philip Appleman (New York: W. W. Norton, 1970) p. 402.

9. Percy Bridgman, *The Logic of Modern Physics* (New York, 1927), p. 83. Quotation taken from Norwood Russell Hanson, *Patterns of Discovery* (Cambridge: University Press, 1972), p. 50.

10. J. O. Westwood, "Mr. Darwin's Theory of Development," *Gardener's Chronicle*, February 11, 1860, p. 122.

SELECTED BIBLIOGRAPHY

Primary Sources

Allen, Grant. "Colour in Nature." Nature 19: 581. 1879.

________. The Colour Sense: Its Origin and Development. An Essay in Comparative Psychology. London: Trubner, 1879.

Argyll, George Douglas Campbell, Duke of. "On Mr. Wallace's Theory of Birds' Nests." Journal of Travel 1: 276-287. 1869.

________. The Reign of Law. 5th ed. New York: George Routledge & Sons, 1869.

Bates, Henry Walter. "Contributions to an Insect Fauna of the Amazon Valley. Coleoptera: Longicornes." Annals and Magazine of Natural History 8: 40-52; 147-152; 212-219; 471-479. 1861.

________. "Contributions to an Insect Fauna of the Amazon Valley. Lepidoptera: Heliconiidae." Transactions of the Linnean Society of London 23: 495-566. 1862.

________. "Contributions to an Insect Fauna of the Amazon Valley. Lepidoptera: Papilionidae." Journal of Entomology, Descriptive and Geographical 1 (4): 218-245. 1861.

________. "Contributions to an Insect Fauna of the Amazon Valley. Part I. Diurnal Lepidoptera." Transactions of the Entomological Society of London 5: 223-228; 335-361. 1858-1861.

________. "Descriptions of Fifty-Two New Species of Phasmidae from the Collection of Mr. W. Wilson Saunders, with Remarks on the Family." Transactions of the Linnean Society of London 25: 321-359. 1865-1866.

________. "Excursion to St. Paulo, Upper Amazons." The Zoologist 16: 6160-6164. 1858.

________. The Naturalist on the River Amazons. A Record of Adventures, Habits of Animals, Sketches of

Brazilian and Indian Life, and Aspects of Nature under the Equator, During Eleven Years of Travel. London: John Murray, 1863.

Bates, Henry Walter. The Naturalist on the River Amazons. A Record of Adventures, Habits of Animals, Sketches of Brazilian and Indian Life, and Aspects of Nature under the Equator, During Eleven Years of Travel. With a Memoir of the Author by Edward Clodd. 2d ed. New York: D. Appleton & Co., 1892.

________. "Notes on South American Butterflies." Transactions of the Entomological Society of London, ser. 2, 5: 1-11. 1858-1861.

________. "Notes upon the Variation of Sexes in Argynnis Diana." Proceedings of the Entomological Society of Philadelphia 4: 204-207. 1865.

Bateson, William. Materials for the Study of Variation Treated with Especial Regard to Discontinuity in the Origin of Species. London: Macmillan, 1894.

Beddard, Frank. Animal Colouration, an Account of the Principle Facts and Theories Relating to the Colours and Markings of Animals. London: Swan Sonnenschein, 1892.

Belt, Thomas. The Naturalist in Nicaragua. London: John Murray, 1874.

Bennett, Alfred W. "Is Protective Mimicry due to Natural Selection?" American Naturalist 11: 3-7. 1877.

________. "Mimicry in Plants." Popular Science Review 9: 1-10. 1872.

________. "On the Theory of Natural Selection Looked at from a Mathematical Point of View. Nature 3: 30-33. 1871.

Bicheno, J. E. "On Systems and Methods in Natural History." Transactions of the Linnean Society of London 15: 479-496. 1827.

Blakiston, Thomas, and Alexander, Thomas. "Protection by Mimicry--A Problem in Mathematical Zoology." Nature 29: 425-426. 1884.

Bree, Charles Robert. An Exposition of Fallacies in the Hypothesis of Mr. Darwin. London: Longman, Green, 1872.

Bree, Charles Robert. Species Not Transmutable, Nor the Result of Secondary Causes. London: Groombridge & Sons, 1860.

Butler, Arthur G. "The Colour Patterns of Butterflies." The Zoologist 3: 1463-1464. 1868.

________. "Mimicry versus Hybridity." Nature 3: 165. 1870.

________. "Remarks upon Certain Caterpillars &c., which are Unpalatable to Their Enemies." Transactions of the Entomological Society of London, 1869, pp. 27-29.

Butler, Joseph. The Analogy of Religion, Natural and Revealed, to the Constitution and Course of Nature. London: Printed for James, John, and Paul Knapton at the Crown in Ludgate St., 1736.

Butler, Samuel. Life and Habit. London: Trubner, 1878.

Carpenter, G. D. Hale. "Attacks of Birds on Butterflies." Transactions of the Entomological Society of London, 1933, pp. 21-26.

________. "Experiments on the Relative Edibility of Insects, with Reference to their Coloration." Transactions of the Entomological Society of London, 1921, pp. 1-105.

________. "Further Evidence that Birds do Attack Butterflies." Proceedings of the Zoological Society of London 3: 223-247. 1937.

Clark, C. A., and Sheppard, P. M. "Interactions Between Major Genes and Polygenes in the Determination of the Mimetic Pattern of Papilio dardanus." Evolution 17: 404-413. 1963.

Collingwood, Cuthbert. "On Homomorphism; or, Organic Representative Form." Liverpool Literary and Philosophical Society, 1860, pp. 181-216.

________. "On Recurrent Animal Form, and its Significance in Systematic Zoology." Annals and Magazine of Natural History 6: 81-91. 1860.

________. Rambles of a Naturalist on the Shores and Waters of the China Sea. London: John Murray, 1868.

[Darwin, Charles.] [Review of] "Contributions to an Insect Fauna of the Amazon Valley." [by Henry Walter Bates] Natural History Review 3: 219-224. 1863.

Darwin, Charles. The Descent of Man and Selection in Relation to Sex. London: John Murray, 1872.

________. A Monograph of the Sub-Class Cirripedia, with Figures of All Species. 2 vols. London: Ray Society, 1851 and 1854.

________. On the Origin of Species. A Facsimile of the First Edition with an Introduction by Ernst Mayr. Cambridge, MA: Harvard University Press, 1966.

________. The Origin of Species. 6th ed. New York: Modern Library, n.d.

________. "The Sexual Colours of Certain Butterflies." Nature 21: 237. 1880.

________. Variation of Animals and Plants under Domestication. 2 vols. London: John Murray, 1868.

________, and Wallace, Alfred Russel. "On the Tendency to Form Varieties; and on the Perpetuation of Varieties and Species by Natural Means of Selection." Proceedings of the Linnean Society of London 3: 45-62. 1858.

Distant, William L. "Difficult Cases of Mimicry." Nature 26: 105. 1882.

________. Rhopalocera Malayana: A Description of the Butterflies of the Malay Peninsula. London: Printed by W. L. Distant, 1882-1886.

________. "The Theory of Mimicry and Mimicking Theories." Annals and Magazine of Natural History, ser. 5, 11: 43-48. 1883.

Dixey, Frederick. "Mimetic Attraction." Transactions of the Entomological Society of London, 1897, pp. 317-332.

________. "On the Phylogeny of the Pierinae, as Illustrated by their Wing-Markings and Geographical Distribution." Transactions of the Entomological Society of London, 1894, pp. 249-334.

________. "On the Relation of Mimetic Patterns to the Original Form." Transactions of the Entomological Society of London, 1896, pp. 65-77.

Doubleday, Edward, and Westwood, John Obadiah. The Genera of Diurnal Lepidoptera: Comprising their Generic

Characters, a Notice of their Habits and Transformations, and a Catalogue of the Species of Each Genus. Illustrated with 86 Plates by William C. Hewitson. 2 vols. London: Longman, Brown, Green, & Longmans, 1846-1852.

Eimer, Gustav Heinrich. *Organic Evolution as the Result of the Inheritance of Acquired Characters According to the Law of Organic Growth*. Translated by J. T. Cunningham. London: Macmillan, 1890.

Eltringham, H. "On the Late Professor Packard's Paper, 'On the Markings of Organisms.'" *Proceedings of the Entomological Society of London*, 1906, pp. xxxvii-xlviii.

Field, George. *Outlines of Analogical Philosophy: Being a Primary View of the Principles, Relations and Purposes of Nature, Science, and Art*. 2 vols. London: Charles Tilt, 1839.

Finn, Frank. "Contributions to the Theory of Warning Colors and Mimicry, No. IV, Summary and Conclusions." *Journal of the Asiatic Society of Bengal* *66*: 613-668. 1897.

Fisher, Ronald. "On Some Objections to Mimicry Theory: Statistical and Genetic." *Transactions of the Entomological Society of London* *75*: 269-278. 1927.

________. *The Genetical Theory of Natural Selection*. Oxford: Clarendon Press, 1930.

Forbes, Edward. "On the Manifestation of Polarity in the Distribution of Organized Beings in Time." *Edinburgh New Philosophical Journal* *52*: 332-337. 1854.

Ford, E. B. *Mimicry*. London: Methuen, 1933.

________. "Problems of Heredity in the Lepidoptera." *Biological Review* *12*: 461-503. 1937.

Goldschmidt, Richard B. "Mimetic Polymorphism, a Controversial Chapter of Darwinism." *Quarterly Review of Biology* *20*: 147-164; 205-250. 1945.

Gray, Asa. *Darwiniana: Essay and Reviews Pertaining to Darwinism*. Cambridge, MA: Harvard University Press, 1963.

Gray, Asa. "Variation and Mimetic Analogy in Lepidoptera." _American Journal of Sciences and Arts_, November, 1863, pp. 285-290.

Gray, John Edward. "Revision of the Species of Lemuroid Animals, with the Description of Some New Species." _Proceedings of the Zoological Society of London_, 1863, pp. 129-152.

________. "Revision of the Species of Trionyclidae Found in Asia and Africa, with the Description of Some New Species." _Proceedings of the Zoological Society of London_, 1864, pp. 76-98.

Haase, Erich. _Untersuchungen über die Mimicry auf Grundlage eines naturlichen Systems der Papilioniden_. Bibliotheca Zoologica, Vol. 8. Stuttgart: Erwin Nägele, 1893.

Hampson, G. F. "Protective and Pseudo-Mimicry." _Nature_ _57_: 364. 1898.

Haughton, Samuel. "On the Form of the Cells Made by Various Wasps and by the Honey Bee; with an Appendix on the Origin of Species." _Annals and Magazine of Natural History_ _11_ (3): 415-428. 1863.

Higgins, Henry H. _Notes by a Field-Naturalist in the Western Tropics_. Liverpool: E. Howell, 1877.

________. "On Colour Patterns in Natural Productions." _Proceedings of the Liverpool Literary and Philosophical Society_, 1857, pp. 132-143.

________. "On the Colour Patterns in Butterflies." _Quarterly Journal of Science_ _5_: 323-329. 1868.

Hingston, R. W. G. _The Meaning of Animal Colour and Adornment. Being a New Explanation of the Colours, Adornments and Courtships of Animals, Their Songs, Moults, Extravagant Weapons, the Differences Between Their Sexes, the Manner of Formation of their Geographic Varieties and Other Allied Problems_. London: Edward Arnold & Co., 1933.

Hubrecht, Ambrosius A. W. "Darwinism _versus_ Wallaceism." _Contemporary Review_ _94_: 629-634. 1908.

Hu ton, Thomas. "On Species and Varieties." _Proceedings of the Entomological Society of London_, 1867, pp. lxxxii-lxxxiii.

Huxley, Thomas Henry. Evidence as to Man's Place in Nature. Ann Arbor Paperbacks. Ann Arbor; University of Michigan Press, 1959.

Kirby, William. "A Description of Some Insects Which Appear to Exemplify Mr. William S. MacLeay's Doctrine of Affinity and Analogy." Transactions of the Linnean Society of London, 1825, pp. 83-110.

________. "Introductory Address, Explanatory of the Views of the Zoological Club, by the Chairman." Zoological Journal 2: 1-8. 1825.

________. Manual of European Butterflies, on the Plan of Stainton's Manual of British Butterflies and Moths. London: Williams & Norgate, 1882.

________. "On the Geographic Distribution of European Rhopalocera." Proceedings of the Entomological Society of London, 1862-64, pp. 481-492.

________. On the Power, Wisdom, and Goodness of God as Manifested in the Creation of Animals and in their History, Habits and Instincts. 2 vols. London: William Pickering, 1833.

________, and Spence, William. An Introduction to Entomology; or Elements of the Natural History of Insects; Comprising an Account of Noxious and Useful Insects, of their Metamorphoses, Food, Stratagem, Habitations, Societies, Motions, Noises, Hibernation, Instinct, etc., etc. 5th ed. 4 vols. London: Printed for Longman, Rees, Orme Brown & Green, 1828.

Lankester, E. Ray. "On the use of the Term Homology in Modern Zoology, and the Distinction between Homogenetic and Homoplastic Agreements." Annals and Magazine of Natural History, ser. 11, 6: 34-43. 1870.

Lowne, B. T. "Observations on Immature Sexuality and Alternate Generation in Insects." Transactions of the Entomological Society of London, 1871, pp. 193-202.

Lubbock, John. "Note on the Colours of British Caterpillars." Transactions of the Entomological Society of London, 1878, pp. 239-243.

Lyell, Charles. The Geological Evidences of the Antiquity of Man, with Remarks on the Theories of the Origin of Species by Variation. London: John Murray, 1863.

Lyell, Charles. Principles of Geology. 3 vols. London: John Murray, 1830-1833.

McAtee, William. Effectiveness in Nature of the So-Called Protective Adaptations in the Animal Kingdom, Chiefly as Illustrated by the Food-habits of Nearctic Birds. Smithsonian Miscellaneous Collections. Vol. 85, No. 7. Washington, D. C.: Smithsonian Institution, 1932.

_______. "The Experimental Method of Testing the Efficiency of Warning and Cryptic Coloration in Protecting Animals from their Enemies." Proceedings of the Academy of Natural Science of Philadelphia, 1912, pp. 281-364.

M'Lachlan, Robert. "Observations on Some Remarkable Varieties of Sterrha Sacria Linn., with General Notes on Variation in Lepidoptera." Transactions of the Entomological Society of London, 1865, pp. 466-4677.

MacLeay, William Sharp. Horae Entomologicae; or, Essays on the Annulose Animals. Vol I, Pts. 1-2. London: Printed for S. Bagster, 1819.

_______. A Letter on the Dying Struggle of the Dichotomous System. London: Printed by Richard Taylor, 1830.

_______. "On Some New Forms of Arachnida." Annals of Natural History 2: 1-14. 1839.

_______. "Remarks on the Identity of Certain Laws Which Have Been Lately Observed to Regulate the Natural Distribution of Insects and Fungi." Transactions of the Linnean Society of London 14: 46-68. 1825.

Marshall, Guy A. K. "Birds a Factor in the Production of Mimetic Resemblances Among Butterflies." Transactions of the Entomological Society of London, 1909, pp. 329-383.

_______. "Five Years Observations and Experiements (1896-1901) on the Bionomics of South African Insects, Chiefly Directed to the Investigation of Mimicry and Warning Colors. Discussion by E. B. Poulton." Transactions of the Entomological Society of London, 1902, pp. 2877-541.

_______. "On Diaposematism With Reference to Some Limitations of the Müllerian Hypothesis of Mimicry." Transactions of the Entomological Society of London, 1908, 93-142.

Meldola, Raphael. "Difficult Cases of Mimicry." Nature 27: 482. 1883.

________. "Mimicry Between Butterflies of Protected Genera." Annals and Magazine of Natural History, ser. 5, 10: 417-425. 1882.

________. "Protective Mimicry." Nature 14: 330. 1876.

________. "Sexual Selection." Nature 3: 508-509. 1871.

Mivart, St. George. "On the use of the Term 'Homology'." Annals and Magazine of Natural History 6: 113-121. 1870.

"Moral Reflections at the Natural History Museum." Punch, May 2, 1906, pp. 312-313.

Morgan, C. Lloyd. Animal Life and Intelligence. London: Edward Arnold, 1890-1891.

Müller, Fritz. "Bemerkenswerthe falle erworbener Aehnlichkeit bei Schmetterlingen." Kosmos 10: 257-267. 1881-1882.

________. "Ituna and Thuridia: A Remarkable Case of Mimicry in Butterflies." Translated and Communicated by Raphael Meldola. Proceedings of the Entomological Society of London, 1879, pp. xx-xxix.

________. "Notes on Brazilian Entomology." Transactions of the Entomological Society of London, 1878, pp. 211-223.

________. "Einige Worte über Leptalis." Jenaische Zeitschrift für Wissenschaft 11: 1-12. 1876.

Murphy, Joseph John. Habit and Intelligence in their Connexion with the Laws of Matter and Force. 2 vols. London: Macmillan, 1869.

Murray, Andrew. The Geographical Distribution of Mammals. London: Day & Son, 1866.

________. "Mimicry and Hybridization." Nature 3: 154-156. 1870.

________. "On Mr. Darwin's Theory of the Origin of Species." Proceedings of the Edinburgh Royal Society 4: 274-292. 1862.

________. "On the Disguises of Nature: Being an Inquiry

into the Laws which Regulate External Form and Colour in Plants and Animals." Edinburgh New Philosophical Journal 11: 66-90. 1860.

Murray, Andrew. "Reply to Mr. Wallace's Theory of Bird's Nests." Journal of Travel 1: 137-145. 1869.

Newbegin, Marion. Colours in Nature. A Study in Biology. London: John Murray, 1898.

Newman, Edward. "Letters on Variation in Lepidoptera. Letter the First--Introduction. Sexual Variation." The Zoologist 2 (2): 721-727. 1867.

Owen, Richard. Lectures on the Comparative Anatomy and Physiology of the Invertebrate Animals. London: Longman, Brown, Green & Longmans, 1843.

Paley, William. Natural Theology; or, Evidences of the Existence and Attributes of the Deity, Collected from the Appearances of Nature. Boston: Richardson, Lord& Holbrook, and Crocker & Brewster, 1831.

Poulton, Edward B. The Colours of Animals. Their Meaning and Use, Especially Considered in the Case of Insects. New York: D. Appleton, 1890.

________. Essays on Evolution. Oxford: Clarendon Press, 1908.

________. "The Experimental Proof of the Protective Value of Colour and Markings in Insects in Reference to Their Vertebrate Enemies." Proceedings of the Zoological Society of London, 1887, pp. 191-274.

________. "Further Experiments upon the Protective Value of Colour and Markings in Insects." Nature 36: 594-595. 1887.

________. "Mimetic Forms of Papilio dardanus (merope) and Acraea johnstoni." Transactions of the Entomological Society of London, 1906, pp. 281-321.

________. "Natural Selection and the Cause of Mimetic Resemblances and Common Warning Colors." Journal of the Linnean Society of London 26: 558-612. 1898.

________. "Termination of the Controversy with Dr. W. L. McAtee." Proceedings of the Entomological Society of London, 1934, pp. 119-120.

Prout, William. Chemistry, Meteorology, and the Function of Digestion Considered with Reference to Natural Theology. Philadelphia: Carey, Lea & Blanchard, 1836.

Ray, John. The Wisdom of God Manifested in the Works of Creation. London: S. Smith, 1704.

Rennie, James. Natural History of Insects. Comprising their Architecture, Transformations, Senses, Food Habits, Collection, Preservation and Arrangement. 3 vols. Boston: Lilly, Watt, Coleman & Holder, 1833.

Riley, Charles V. "Mimicry as Illustrated by These Two Butterflies, with some Remarks on the Theory of Natural Selection." Third Annual Report on the Noxious, Beneficial and Other Insects of the State of Missouri. Jefferson City, Mo.: Horace Wilcox, 1871.

Roget, Peter Mark. Animal and Vegetable Physiology Considered With Reference to Natural Theology. 2 vols. Philadelphia: Carey, Lea & Blanchard, 1836.

Romanes, George J. Darwin and After Darwin. An Exposition of the Darwinian Theory and a Discussion of Post-Darwinian Questions. 3 vols. Chicago: Open Court Publishing Company, 1892-1897.

________. "Mr. Wallace on Darwinism." Contemporary Review 56: 244-258. 1889.

Schmidt, Oscar. The Doctrine of Descent and Darwinism. New York: D. Appleton & Co., 1875.

Schouw, Joachim. The Earth, Plants, and Man: Popular Pictures of Nature. Translated and Edited by Arthur Henfrey. London: Henry G. Bohn, 1852.

Semper, Karl. The Natural Conditions of Existence as they Affect Animal Life. 3rd ed. London: Kegan, Paul Trench & Co., 1853.

Shull, A. Franklin. "The Needs of the Mimicry Theory." Science 85: 496-498. 1937.

Slater, J. W. "On the Food of Gaily-Coloured Caterpillars." Transactions of the Entomological Society of London, 1877, pp. 205-209.

Smith, Frederick. "On the Construction of Hexagonal Cells by Bees and Wasps." Transactions of the Entomological Society of London, 1864, pp. 131-141.

Stainton, Henry Tibbets. "How may the Onward Progress of the Study of Entomology be Best Furthered?" Transactions of the Entomological Society of London 4(2): 138-142. 1856-1858.

Strickland, Hugh. "On the Structural Relations of Organized Beings." Philosophical Magazine, ser. 3, 28: 354-364. 1846.

________. "On the True Method of Discovering the Natural System in Zoology and Botany." Annals and Magazine of Natural History 6: 184-194. 1841.

Thayer, A. H. "Protective Coloration in its Relation to Mimicry, Common Warning Colours, and Sexual Selection." Transactions of the Royal Entomological Society of London, 1903, p. 553.

Thayer, Gerald H. Concealing Coloration in the Animal Kingdom. London: Macmillan & Co., 1909.

Trimen, Roland. "On Some Remarkable Mimetic Analogies Among African Butterflies." Transactions of the Linnean Society of London 26: 497-523. 1870.

Tylor, Alfred. Coloration in Animals and Plants. London: Alabaster, Passmore & Sons, 1886.

VanDyck, W. "On the Modification of a Race of Syrian Street Dogs by Means of Sexual Selection. With a Preliminary Notice by Charles Darwin." Proceedings of the Zoological Society of London 25: 367-370. 1882.

Wallace, Alfred Russel. "Colour in Nature." Nature 19: 501-505. 1879.

________. [Review of] "The Colours of Animals." [by Edward Poulton] Nature 42: 437. 1890.

________. "The Colours of Animals and Plants." American Naturalist 11: 641-662; 713-728. 1877.

________. Contributions to the Theory of Natural Selection. New York: Macmillan, 1871.

________. Darwinism. New York: Humboldt Publishing Co., 1889.

________. "Darwinism versus Wallaceism." Contemporary Review 94: 716-717. 1908.

Wallace, Alfred Russel. [Review of] "The Descent of Man and Selection in Relation to Sex." [by Charles Darwin] Academy 2: 177-183. 1871.

________. "Dr. Fritz Müller on Some Difficult Cases of Mimicry." Nature 26: 36-37. 1882.

________. "Exposition of the Theory of Mimicry of Adaptive Resemblances as Explaining Anomalies of Sexual Variation." Proceedings of the Entomological Society of London, 1866, pp. 38-41.

________. The Geographical Distribution of Animals. 2 vols. New York: Harper & Bros., 1876.

________. The Malay Archipelago: The Land of the Orang-utan, and the Bird of Paradise. A Narrative of Travel, with Studies of Man and Nature. London: Macmillan, 1869.

________. "Mimicry Versus Hybridity." Nature 3: 165-166. 1870.

________. A Narrative of Travels on the Amazon and Rio Negro, With an Account of the Native Tribes, and Observations on the Climate, Geology, and Natural History of the Amazon Valley. London: Reeve & Co., 1853.

________. A Narrative of Travels on the Amazon and Rio Negro. 2d ed. of 1889. With a new Introduction by H. Lewis McKinney. New York: Dover Publications, Inc., 1972.

________. Natural Selection and Tropical Nature: Essays on Descriptive and Theoretical Biology. London: Macmillan, 1891.

________. "Note on the Theory of Permanent and Geographical Varieties." The Zoologist 16: 5887-5888. 1858.

________. "On the Law Which has Regulated the Introduction of New Species." Annals and Magazine of Natural History 16: 184-196. 1855.

________. "On the Phenomena of Variation and Geographical Distribution, as Illustrated by the Papilionidae of the Malayan Region." Transactions of the Linnean Society of London 25: 1-72. 1865.

________. "Philosophy of Birds' Nests." Intellectual Observer 11: 413-420. 1867.

Wallace, Alfred Russel. "Remarks on the Rev. S. Haughton's Paper on the Bee's Cell, and on the Origin of Species." Annals and Magazine of Natural History 12 (3): 303-308. 1863.

________. Tropical Nature and Other Essays. London: Macmillan, 1878.

Walsh, Benjamin D. "On Certain Entomological Speculations of the New England School of Entomologists." Proceedings of the Entomological Society of Philadelphia, 1864, pp. 207-249.

________, and Riley, Charles V. "Imitative Butterflies." The American Entomologist 1: 189-193, 1869.

Weir, John Jenner. "Further Observations on the Relation Between the Colour and the Edibility of Lepidoptera and Their Larvae." Transactions of the Entomological Society of London, 1870, pp. 337-339.

________. "On Insects and Insectivorous Birds; and Especially on the Relation Between the Colour and the Edibility of Lepidoptera and Their Larvae." Transactions of the Entomological Society of London, 1869, pp 21-26.

Weismann, August. The Evolution Theory. Translated by J. Arthur Thomson and Margaret R. Thomson. 2 vols. London: Edward Arnold, 1904.

________. Studies in the Theory of Descent. With Notes and Additions by the Author. Translated and edited by Raphael Meldola, with a prefatory notice by Charles Darwin. London: Sampson Low, Marston, Searle, & Rivington, 1882.

Westwood, John Obadiah. The Entomologist's Textbook: An Introduction to the Natural History, Structure, Physiology, and Classification of Insects, Including the Crustacea and Arachnida. London: Willim S. Orr & Co., 1838.

________. "Illustrations of the Relationships Existing Amongst Natural Objects, Usually Termed Affinity and Analogy, Selected from the Class of Insects." Transactions of the Linnean Society of London 18: 409-422, 1841.

________. "Mr. Darwin's Theory of Development." Annals and Magazine of Natural History 5 (3): 347-348. 1860.

Whewell, William. Astronomy and General Physics, Considered With Reference to Natural Theology. London: William Pickering, 1833.

Wollaston, Thomas Vernon. Coleoptera Atlantidum, Being an Enumeration of the Coleopterous Insects of the Madeiras, Salvages, and Canaries. London: John Van Voorst, 1865.

________. [Review of] "On the Origin of Species by Means of Natural Selection; or, the Preservation of Favoured Races in the Struggle for Life." [by Charles Darwin] Annals and Magazine of Natural History 5: 132-143. 1860.

Secondary Sources

Barlow, Nora, ed. The Autobiography of Charles Darwin, 1809-1882. New York: W. W. Norton, 1969.

Barzun, Jacques. Darwin, Marx, Wagner. 2d ed. Garden City, New Jersey: Doubleday & Co., 1958.

Beddall, Barbara G., ed. Wallace and Bates in the Tropics. London: Macmillan, 1969.

________. "Wallace, Darwin and the Theory of Natural Selection." Journal of the History of Biology 1: 261-323, 1968.

Boas, George. The History of Ideas: An Introduction. New York: Charles Scribner's Sons, 1969.

Boyden, Alan. "Homology and Analogy: A Century After the Definitions of 'Homologue' and 'Analogue' by Richard Owen." Quarterly Review of Biology 18: 228-241, 1943.

________. "Homology and Analogy. A Critical Review of the Meaning and Implications of These Concepts in Biology." American Midland Naturalist 37: 648-669. 1947.

Burkhardt, Frederick. "England and Scotland. The Learned Societies." The Comparative Reception of Darwinism. Edited by Thomas Glick. Austin: University of Texas Press, 1974.

Burrow, J. W. "Evolution and Anthropology in the 1860's: The Anthropological Society of London, 1863-71." Victorian Studies 7: 139-154. 1963.

Campbell, Bernard, ed. Sexual Selection and the Descent of Man. Chicago: Aldine, 1972.

Cannon, Walter B. "The Bases of Darwin's Achievement: A Revaluation." Victorian Studies 5: 109-134. 1961.

________. "The Problem of Miracles in the 1830's." Victorian Studies 4: 4-32. 1960.

Cott, Hugh B. Adaptive Coloration in Animals. London: Methuen & Co., 1940.

Darlington, C. D. "The Origin of Darwinism." Scientific American 200: 60-66. 1956.

Darwin, Francis, ed. The Life and Letters of Charles Darwin. 3 vols. London: John Murray, 1888.

________, and Seward, A. C., eds. More Letters of Darwin. 2 vols. New York: D. Appleton, 1903.

Dewey, John. The Influence of Darwinian Philosophy and Other Essays in Contemporary Thought. New York: H. Holt & Co., 1910.

De Ruiter, L. "Some Remarks on Problems of the Ecology and Evolution of Mimicry." Archives Neerlandaises de Zoologie 13 (suppl.): 351-368. 1958.

Dewar, Douglas, and Finn, Frank. The Making of Species. London: John Lane, 1909.

Ellegård, Alvar. Darwin and the General Reader: The Reception of Darwin's Theory of Evolution in the British Periodical Press, 1859-1872. Gothenburg Studies in English, Vol. 8. Gothenburg: Elanders, 1958.

________. "The Darwinian Theory and Nineteenth-Century Philosophies of Science." Journal of the History of Ideas 18: 362-393, 1957.

________. "The Darwinian Theory and the Argument from Design." Lychnos, 1956, pp. 173-192.

Evans, Mary Alice. "Mimicry and the Darwinian Heritage." Journal of the History of Ideas 26: 211-220, 1965.

Ferré, Frederick. Language, Logic and God. New York: Harper & Row, 1969.

Fleming, Donald. "Charles Darwin, the Anaesthetic Man." Victorian Studies 4: 219-234. 1961.

Ford, E. B. "The Genetic Basis of Adaptation." Evolution. Edited by Gavin De Beer. Oxford: Clarendon Press, 1938.

Freeman, John. The Life of the Rev. William Kirby, M.A. London: Longman, Brown, Green & Longmans, 1852.

Garfinkle, Norton. "Science and Religion in England, 1790-1800: The Critical Response to the Work of Erasmus Darwin." Journal of the History of Ideas 16: 376-388,. 1955.

George, Wilma. Biologist Philosopher. A Study of the Life and Writings of Alfred Russel Wallace. New York: Abelard-Schuman, 1964.

Ghiselin, Michael. The Economy of Nature and the Evolution of Sex. Berkeley: University of California Press, 1974.

________. The Triumph of the Darwinian Method. Berkeley: University of California Press, 1969.

Gillispie, Charles C. The Edge of Objectivity. An Essay on the History of Scientific Ideas. Princeton, New Jersey: Princeton University Press, 1960.

Glass, Bentley; Temkin, Owsei; and Straus, William L., Jr., eds. Forerunners of Darwin: 1745-1859. Baltimore: Johns Hopkins University Press, 1959.

Glick, Thomas, ed. The Comparative Reception of Darwinism. Austin: University of Texas Press, 1974.

Gray, Jane Loring, ed. Letters of Asa Gray. Boston: Houghton, Mifflin & Co., 1893.

Greene, John. "The Kuhnian Paradigm and the Darwinian Revolution in Natural History." Perspectives in the History of Science and Technology. Edited by Duane H. D. Roller. Norman: Oklahoma University Press, 1971.

Gruber, Howard E. Darwin on Man: A Psychological Study of Scientific Creativity. With Darwin's early and unpublished notebooks, transcribed and annotated by Paul H. Barrett. London: Dutton, 1974.

Gundry, D. W. "The Bridgewater Treatises and their Authors." History 31: 140-152. 1946.

Gunther, Albert E. A Century of Zoology at the British Museum Through the Lives of Two Keepers, 1815-1914. New York: Science History Publications, 1975.

Haas, Otto, and Simpson, George G. "Analysis of Some Phylogenetic Terms, with Attempts at Redefinition." Proceedings of the American Philosophical Society 90: 319-347. 1946.

Hanson, Norwood Russell. Patterns of Discovery. Cambridge: University Press, 1972.

Harris, R. W. Reason and Nature in 18th-Century Thought. London: Barnes & Noble, 1969.

Henderson, Gerald M. "Alfred Russel Wallace: His Role and Influence in Nineteenth Century Evolutionary Thought." Unpublished Ph. D. dissertation, University of Pennsylvania, 1960.

Himmelfarb, Gertrude. Darwin and the Darwinian Revolution. The Norton Library. New York: W. W. Norton, 1968.

________. Victorian Minds. Harper Torchbooks. New York: Harper & Row, 1970.

Houghton, Walter E. The Victorian Frame of Mind. New Haven: Yale University Press, 1957.

Huxley, Julian. "Darwin's Theory of Sexual Selection and the Data Subsumed by it, in the Light of Recent Research." American Naturalist 72: 416-433. 1938.

________. "The Emergence of Darwinism." Journal of the Linnean Society (Zoology) 44: 1-73. 1958.

________. Evolution: The Modern Synthesis. New York: John Wiley, 1964.

________. "The Present Standing of the Theory of Sexual Selection." Evolution. Edited by Gavin De Beer. Oxford: Clarendon Press, 1938.

Irvine, William. Apes, Angels, and Victorians. Meridian Books. New York: World Publishing Co., 1970.

Jones, W. P. "The Vogue of Natural History in England, 1750-1770. Annals of Science 2: 345-352. 1937.

Kettlewell, H. B. D. "Darwin's Missing Evidence." Scientific American 200: 48-53. 1959.

Lack, David. Life of the Robin. London: H. F. & G. Witherby, 1944.

Lovejoy, Arthur. The Great Chain of Being. A Study of the History of an Idea. Harper Torchbooks. New York: Harper & Row, 1960.

McKinney, Henry Lewis. "Alfred Russel Wallace and the Discovery of Natural Selection." Journal of the History of Medicine and Allied Sciences 21: 333-357. 1966.

________. "Wallace's Earliest Observations on Evolution: 28 December 1845." Isis 60: 370-373. 1969.

________. Wallace and Natural Selection. New Haven: Yale University Press, 1972.

Marchant, James, ed. Alfred Russel Wallace: Letters and Reminiscences. 2 vols. London: Cassel & Co., 1916.

Mayr, Ernst. Systematics and the Origin of Species. New York: Columbia University Press, 1942.

Meyerriecks, Andrew J. Man and Birds: Evolution and Behavior. New York: Bobbs-Merrill, 1972.

Millhauser, Milton. Just Before Darwin: Robert Chambers and Vestiges. Middletown, Connecticut: Wesleyan University Press, 1959.

Neave, S. A. The Centenary History of the Entomological Society of London, 1833-1933. London: Richard Clay & Sons, 1933.

Passmore, John. "Darwin's Impact on British Metaphysics." Victorian Studies 3: 19-40, 1959.

Peckham, Morse, ed. The Origin of Species, by Charles Darwin. A Variorum Text. Philadelphia: University of Pennsylvania Press, 1959.

Poulton, Edward Bagnall. "The Influence of Darwin upon Entomology." Entomologist's Record and Journal of Variation, 1901, pp. 72-76.

Remington, Charles L. "Historical Backgrounds of Mimicry." Proceedings of the XVth International Congress of Zoology 4: 143-149. 1963.

Remington, Jeanne E., and Remington, Charles. "Mimicry, a Test of Evolutionary Theory." Yale Scientific Magazine, October, 1957, pp. 10-21.

Roller, Duane H. D., ed. Perspectives in the History of Science and Technology. Norman: Oklahama University Press, 1971.

Smith, Ray; Mittler, Thomas; and Smith, Carroll, eds. History of Entomology. Palo Alto, Ca.: Annual Reviews, 1973.

Stecher, Robert M. "The Darwin-Bates Letters, Correspondence Between Two Nineteenth-Century Travellers and Naturalists." Annals of Science 25: 1047; 95-115. 1969.

Stephen, Leslie. History of English Thought in the 18th Century. 2 vols. Harbinger Books. New York: Harcourt Brace & World, 1962.

Wade, J. S. "Bibliography of Entomologists." Annals of the Entomological Society of America 21: 489-520. 1928.

Wallace, Alfred Russel. My Life. A Record of Events and Opinions. 2 vols. New York: Dodd, Mead & Co. 1905.

Wickler, Wolfgang. Mimicry in Plants and Animals. Translated by R. D. Martin. World University Library. New York: McGraw-Hill Book Company, 1968.

Woodcock, George. Henry Walter Bates, Naturalist of the Amazons. London: Faber and Faber, 1969.